Some Advance Comments on *HANDBOOK OF CUBIK MATH*

There is no more enjoyable way to learn the basics of group theory than to study this book carefully with (of course) Rubik's Cube in your hands. It is by all odds the best book yet written, or likely to be written, on the wonders and the dark, unsolved mysteries of the cube. Think not that the cube is just a toy. As Frey and Singmaster make clear, it is one of the most amazing teaching aids in the entire history of mathematics.

—Martin Gardner, author and editor.

This handbook gives a fascinating account of the relationship between the Rubik's Cube puzzle and an area of higher mathematics known as the theory of finite groups. The discussion is careful and lucid and should be accessible to anyone willing to think carefully about the cube. It should be particularly useful for high-school students interested in the cube and also as a supplement for a college course covering finite groups. There are a broad range of exercises and various facts about the cube that have not appeared in detail in print before.

—Dr. Joe P. Buhler, Professor of Mathematics, Reed College, Oregon, and Professor of Mathematics, Pennsylvania State University. (He was interviewed on NBC-TV's "20/20" program about the cube.)

Frey and Singmaster give a nicely blended discussion of specific algorithms for the cube together with some underlying theoretical concepts from group theory. Thus, cubists will be able to cure their insomnia by reading chapter 3, and then perhaps satisfy their aroused curiosity about what is really going on by reading further.

There is no doubt that the cube is far and away the most interesting concrete example of a finite group in existence. As such, it presents a great pedagogical opportunity. While finite group theory may not become a household word as a result, books like this one will help demystify a previously rather obscure subject.

—David M. Goldschmidt, Professor of Mathematics, University of California, Berkeley, and group-theory editor of PROCEEDINGS OF THE AMERICAN MATHEMATICAL SOCIETY.

This is the sort of book you could give to brighter students to have them discover the group properties of the cube on their own. The book is clearly and simply written.

—Leroy Sachs, classroom teacher, Clayton, Missouri.

Handbook of
Cubik Math

by
Alexander H. Frey, Jr.
and
David Singmaster

ENSLOW PUBLISHERS
Bloy Street and Ramsey Avenue
Box 777
Hillside, New Jersey 07205

Library of Congress Cataloging in Publication Data:

Frey, Alexander H.
  Handbook of cubik math.

  Bibliography: p.
  Includes index.
  1. Rubik's Cube. 2. Groups, Theory of.
I. Singmaster, David. II. Title.

QA491.F73	512'.22	81-12525

ISBN 0-89490-060-9   AACR2
ISBN 0-89490-058-7 (pbk.)

Printed in the United States of America

10 9 8 7 6 5 4 3 2 1

ACKNOWLEDGMENTS

The works of many people over the last decade have contributed to this book. Not only did Ernö Rubik devise the Magic Cube, but he also has contributed substantially to an understanding of the possible movements of cube pieces. Both John Conway and Tamás Varga are among the early pioneers in the study of cubik math, and to these and many others like them, the authors are particularly indebted.

Much of the material in this book is derived from David Singmaster's *Notes on Rubik's Magic Cube*. Related to this, there have been literally thousands of letters from many correspondents over the last several years that have contributed to both books. To facilitate the further exchange of new ideas, there is now a cubik circular. Persons may write to D.S. Ltd., 66 Mount View Rd., London N4 4JR, United Kingdom for particulars. Interesting mathematical puzzles will also be distributed by this company.

The first U.S. importer of the Magic Cube was Logical Games Inc., Haymarket, Virginia. We are most grateful to its founder, Bela Szalai, whose idea for this book brought the authors together.

Many people have contributed to the preparation of the book. Lowell Smith and Alice H. Frey both read the entire manuscript, eliminating many errors. Cathy Schrott of Jantron Inc. did an outstanding job of page layout and typesetting. Special appreciation is extended to Helen Koehler Frey who not only read the entire manuscript several times, but also supplied the patient encouragement needed to finally complete this project.

Alexander H. Frey, Jr.
and
David Singmaster

PREFACE

Handbook of Cubik Math is a book about problem solving and some of the fundamental techniques used in problem solving throughout mathematics and science. Both the problems and the illustrations of concepts for solving them are drawn from Rubik's Magic Cube.

Ernö Rubik invented the Magic Cube as an aid to developing three-dimensional skills in his students. Little did he realize the impact that this puzzle would come to have. In 1980 alone, approximately five million cubes were sold. Predictions for future years are that sales will continue at more than twice that rate. In almost every neighborhood children — and adults — are playing with the cube.

It certainly enhances three dimensional thinking. However, even greater educational value has been found by mathematicians. For them the cube gives a unique physical embodiment of many abstract concepts which otherwise must be presented with only trivial or theoretical examples. Cube processes are non-commutative — that is, changing the order of movements produces different results. Cube processes generate permutations of the pieces of the cube. Sometimes different processes generate the same permutation so that, by looking at the cube, you cannot tell which process was used. This defines an equivalence between processes. The concepts of an identity process, inverses, the cyclic order of a process, commutators, and conjugates all play a part in solving problems on the cube. By experimenting on the cube, a student learns about these concepts and their relation to problem defining and problem solving without having to rely solely on his faith in the teacher or the text.

Perhaps surprisingly, one of the most fundamental concepts which cubik math illustrates is the use of symbolic notation. It is extremely rare to find anyone who can master the complexities of the cube without writing down what movements he has made or is planning to make. Without a good symbolic notation this is cumbersome at best. For communicating about the cube with others a common notation is mandatory.

The *Handbook of Cubik Math* in the early technical chapters orients the reader to the basic problem of the cube. It introduces a standard notation — one which is internationally accepted. Then it describes a logical method for restoring any scrambled cube to its pristine state where every face is a solid color. No background of complex or sophisticated mathematical concepts is required in these first three chapters. Many good students in their early teens have mastered these ideas. At the end of Chapter 3, several games are introduced. Playing these will enhance the competitor's understanding of the concepts inherent in controlled modification of the state of the cube.

One might think that after learning how to solve the cube — that is, how to restore it to its monochromatic-sided state — a person would lose interest in the cube. We thought so before we had taught many people how to solve it, only to find that with their increased understanding came increased curiosity. They wanted to understand more about how the cube worked, why processes produced the results they do, and what they could do to enhance their mastery of the cube.

Seldom does one realize at this point that the concepts which appeared so logical for solving the cube problem are, in fact, the concepts of identities, inverses, commutators, and conjugates. Chapter 4 defines these generalized concepts with many examples and exercises from the cube. These principles are applied to derive new techniques for manipulating the cube. Then in Chapter 5 these improvements are applied to obtain better ways to restore the cube.

It is in Chapter 6 that the mathematical concepts become more sophisticated. It is here that the concept of a group is introduced. The structure and the size of the cube group and its subgroups are explored in Chapters 6 and 7. This leads finally to a discussion of normal subgroups and the isomorphisms of subgroups and factor groups in Chapter 8.

It is expected that some students of the cube will only be ready to absorb material through Chapter 3. Others will be able to work through Chapter 5. The more advanced students will work all the way through to the end. At all stages it is necessary to have easy access to a cube. The cube is the best teacher and experimentation is the best learning technique.

CONTENTS

CHAPTER 1

INTRODUCTION

The Magic Cube or Rubik's Cube is an ingenious puzzle invented by Ernö Rubik, a sculptor, architect, and teacher of three dimensional design at the Academy of Applied Art in Budapest.

When new, it looks like a cube, about the size of a fist, with each face colored with one of six bright colors. Closer examination shows that the cube is divided in three along each direction so that it appears to be a 3 x 3 x 3 array of little cubes — called *cubies*. Thus each face of the cube is really a 3 x 3 pattern of little faces — called *facelets* — of the small cubies.

One of the first questions about the cube which we usually are asked is, "What is the problem?". We explain that the problem is to devise a method by which, starting with a randomly scrambled cube, you can restore the cube to the position where each face has a single solid color. About half of the people then respond by saying "Oh, so I am supposed to figure out how to take it apart!" We say, "No, the cube does not come apart. At least it is not supposed to."

About half of those people then go away to work out how the cube comes apart, muttering something like, "These fellows are no help. They clearly don't understand the problem." But, for the rest of you who are still here, we can go on to the usual next question. "What movements can the cube make?" One could answer that each of the six faces can be rotated about its central cubie as shown in Figure 1-1. After turning any one of the faces you can now turn any other face. This causes the colored facelets to move about. Sounds simple, doesn't it? If we stop there, we have made the problem about as difficult as one possibly could. Why? Because we may have turned off the most fruitful line of inquiry leading to solutions of the problem. It is important to understand a great deal more about how the cube moves than just that each of the six faces can be rotated.

>   *WARNING:* It only takes a few random turns to thoroughly confuse your cube! Each face soon looks like a Mondrian painting. Without a solution, such as that given in Chapter 3, it could take you a long time to restore your cube.

Masochists who insisted on restoring their cubes without any help have taken weeks or months. Several of our friends took nine months to a year! When you understand the basic strategy taught in this book, you will be able to restore any scrambled cube without referring again to the book. The strategy does not require you to memorize any sequences of moves. You are taught the reason for each and every face turn.

CUBE MOVEMENTS

Any of the six faces of the cube can be rotated.

Figure 1-1

CHAPTER 2

A CUBIK ORIENTATION

Before developing a strategy for restoring the cube, it helps to study the cube a little while. What can we observe about the cube that may help us with the solution? What simple terminology and notation will describe the pieces and movements of the cube? Figure 2-1 gives a summary of the terminology and notation to be developed in this chapter.

1. CUBIES AND CUBICLES

Looking at the cube as a whole, at first glance, it appears to be made up of 3 x 3 x 3 = 27 cubies in three layers, each layer being a three-by-three square of small cubies. However, it is only possible to see the outside of the cube, so that only 26 cubies can be seen. The one in the center is only imaginary. Also, all that we can see of each of the 26 visible cubies are the colored facelets which combine to form the six faces of the entire cube. Each face of the cube is made up of nine such facelets. Thus there are 6 x 9 = 54 facelets on the cube.

3

SUMMARY OF TERMINOLOGY AND NOTATION

Terminology	Definition or Abbreviation
Cubies	The small cube pieces which make up the whole cube.
Cubicles	The spaces occupied by cubies.
Facelets	The faces of a cubie.
Types of Cubies: Corner, Edge, and Center	A corner cubie has three facelets. An edge cubie has two facelets. A center cubie has one facelet.
Home Location — of a cubie	The cubicle to which a cubie should be restored.
Home Position — of a cubie	The orientation in the home location to which a cubie should be restored.
Positional Names for Cube Faces	Up Down Right Left Front Back
Notation for Cubicles — shown in *italics*	Lower case initials. For example, *uf* denotes the Up-Front edge cubicle.
Notation for Cubies — shown in *italics*	Upper case initials. For example, *URF* denotes the cubie whose home position is in the Up-Right-Front corner.
Notation for Face Turns — shown in BLOCK CAPITAL LETTERS	The initials, U, F, R, D, B, and L denote clockwise quarter turns. $U^{-1}, F^{-1}, R^{-1}, D^{-1}, B^{-1}$, and L^{-1} denote counter-clockwise quarter turns. U^2, F^2, R^2, D^2, B^2, and L^2 denote half turns.
Moving the Whole Cube	*U, F, R, D, B,* and *L* denote clockwise turns of the whole cube behind the indicated face.

Figure 2-1

Look now at the cubies which make up the cube. Notice that the cube has three types of cubies. Some cubies have three visible facelets as indicated in Figure 2-2. These are called *corner pieces*. There are eight corner pieces corresponding to the eight corners of the cube. Other cubies have only two visible facelets as indicated in Figure 2-3. These cubies fill in the space along an edge between two corner pieces. Therefore, they are called *edge pieces*. There are twelve edge pieces, one on each of the twelve edges of the cube. The third type of cubie has only one visible facelet. This facelet, as shown in Figure 2-4 is in the middle of a face. Thus these cubies are called *center pieces*. There are six center pieces corresponding to the six faces of the cube.

By rotating different faces of the cube, the cubies can be moved about. Each cubie moves to the location vacated by another cubie. These locations are called *cubicles*. The locations occupied by corner cubies are corner cubicles and the locations occupies by edge cubies are edge cubicles. Observe that no matter how faces are rotated, the corner pieces always move from one corner cubicle to another corner cubicle and the edge pieces always move from one edge cubicle to another edge cubicle. Rotating a face never moves a center cubie from one face to another. The center pieces have a fixed location relative to the other center pieces. They can only be spun in place. This is a particularly important observation, because it shows the following:

> *The color of the center piece of any face defines the only color to which that face of the cube can be restored.*

For each center piece the color of the opposite center piece never changes. Furthermore, if two opposite center pieces are placed in the positions of north and south poles respectively, then the sequential order of the other four center pieces around the equator is always the same.

THE CORNER CUBIES,
SHADED, HAVE
THREE FACELETS

THE EDGE CUBIES,
SHADED, HAVE
TWO FACELETS

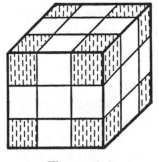

Figure 2-2

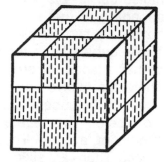

Figure 2-3

THE CENTER CUBIES, SHADED, HAVE ONE FACELET

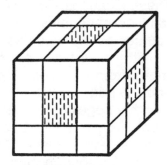

Figure 2-4

Since the center cubie of each face determines the only color to which that face can be restored, we can also define the one and only cubicle in which each cubie can be placed to restore the cube. For example, if the two facelets of an edge cubie are orange and green, then that piece must be placed in the unique edge cubicle between the orange center piece and the green center piece as shown shaded in Figure 2-5. Furthermore, the cubie must be

AN EDGE HOME POSITION

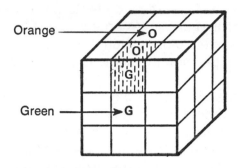

Figure 2-5

placed in that cubicle so that its orange facelet is next to the orange center piece and the green facelet is next to the green center piece.

Similarly, if the three facelets of a corner cubie are orange, green, and white then, to restore that cubie, it must be placed in the corner cubicle where the orange face, the green face, and the white face meet — shaded in Figure 2-6. Furthermore, its orange, green, and white facelets must be on the orange, green, and white faces respectively.

A CORNER HOME POSITION

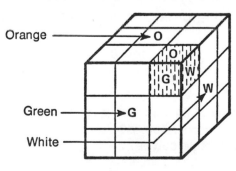

Figure 2-6

For each edge and corner cubie in the cube, the unique cubicle to which it must be restored is called the *home location* for that cubie. When a cubie is in its home location and its facelet colors match the colors of the center pieces on each face, then the cubie is said to be in its *home position*.

It is possible for a cubie to be in the cubicle of its home location without being in its home position. A corner piece in this condition is said to be *twisted* in its home location. An edge piece in this condition is said to be *flipped* in its home location. Figure 2-7 shows a twisted corner cubie and a flipped edge cubie. Thus, each corner and edge cubie has a unique home location and in that cubicle it has a unique placement which puts it in its home position.

TWISTED AND FLIPPED CUBIES

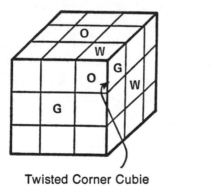 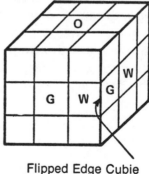

Twisted Corner Cubie   Flipped Edge Cubie

Figure 2-7

EXERCISES:

2.1-1 How many of the 54 facelets of the cube are
    a. facelets of corner cubies?
    b. facelets of edge cubies?
    c. facelets of center cubies?

2.1-2 At how many locations can an edge cubie be placed so that the colors of both of the two adjacent center cubies are different from both colors on the facelets of that edge cubie?

2.1-3   In what cubicle can a corner cubie be placed so that none of the center cubies adjacent to that cubicle has the color of any of the three facelets of that corner cubie? Describe the cubicle location relative to the home position of the cubie.

2.   ORIENTATION BASED ON THE CENTER FACELETS

To discuss the movements of the cube and its cubies, we need to establish a terminology and notation. The most important quality of any terminology and notation is that it be accepted and used in the same way by all the people who need to communicate about the subject. There is one terminology and notation that has been accepted internationally by most students of the cube. It was devised by David Singmaster of Polytechnic of the South Bank, London, England. That terminology is the one we use in this book.

Many people who work with the cube have developed their own terminology and notation for cube movements. Perhaps you already have a notation of your own. Different people's notations reflect their diverse ways of approaching the cube problems. Even if one of these were better than the one we use in this book, we would not want to change because the one presented in this book is already the most widely used.

We want to be able to discuss how cubies move from cubicle to cubicle in the cube. To do this, we need to describe the location of each cubicle. Since rotating a face never changes the location of the center cubies, it is natural to use these center pieces as reference points to describe the locations of the cubicles.

It is tempting to name each face by the color of the center facelet. However, different manufacturers use different colors and even the same manufacturer does not always keep the same relative placement for coloring the faces of different cubes. In discussions among people with such different cubes, using colors to identify cubies, cubicles, and the movements of the cube may be more confusing than helpful.

To avoid the confusions caused by different cube colorings, each of the faces is named based on its position relative to the person holding the cube. Singmaster has chosen the following six names for the six faces: Front, Back, Right, Left, Up, and Down. These faces are designated by their initials as follows:

$$
\begin{aligned}
\text{Front} &= \text{F} \\
\text{Back} &= \text{B} \\
\text{Right} &= \text{R} \\
\text{Left} &= \text{L} \\
\text{Up} &= \text{U} \\
\text{Down} &= \text{D}
\end{aligned}
$$

It is very convenient to be able to abbreviate the names of the faces by their initials. Singmaster chose these names to avoid the ambiguities presented by the initials of some of the logical English words for the faces; Right/Rear and Back/Bottom.

To designate a face on your own cube as the Up face, choose any center cubie which you like to be the Up-face center piece. From then on the pattern of all facelets which appear at any moment on the same face as that chosen center-piece facelet will constitute the *Up face*. The Up *layer* is the set of all cubies which have a facelet on the Up face. The color of the Up-face center facelet is the color to which the Up face must be restored.

After choosing a center cubie for the Up face, you can choose any of four other center cubies for the Front-face center piece. After you have chosen colors for the center facelets of the Up face and the Front face, then all the other center facelet colors are fixed by the way the manufacturer put the colors on the cube. Thus we now have matched each of the six faces of the cube with a different one of the six positional names Up, Front, Right, Down, Back, and Left.

This correspondence between the cube faces and positions in space defines an orientation of the cube. The *orientation* of an object is its position relative to an agreed-upon point of reference. In many cases the point of reference may be moveable. For example, we can use North or the "front" of a car or even the direction in which you are looking as points of reference and describe the orientation of other objects accordingly. Thus if a car has turned over in an accident we say that the part of the car which is on the bottom is the "top" of the car. The same is true after we have defined an orientation for the cube. We may turn the cube over to look at the Down face. This may temporarily put the Up face on the bottom and the Front face in the back. But this does not change the orientation of the cube. The color which we assigned to the Up face remains the Up-face color and the color which we assigned to the Front is still the Front-face color. However, we can reorient the cube. We say that we have reoriented the cube whenever we assign a different color to one of the six positional names. Thus if you turn the cube around just to look at the back then return it, that does not reorient the cube. But, we will find times when it is useful to turn it around and keep the color that was the Back-face color in front and call it the Front-face color. That re-naming process is called *reorienting* the cube.

We also use the positional names of the faces to identify the cubicles, cubies, and facelets. For example, the edge cubicle between the Up-face center piece and the Front-face center piece can be called either the Up-Front cubicle or the Front-Up cubicle. The edge cubie which has one Down-colored facelet and one Right-colored facelet is the Down-Right cubie — or also the Right-Down cubie. The home location of the Left-Back-Up corner cubie is the corner cubicle where the Up face, the Left face and the Back face meet — the Up-Left-Back cubicle. This cubie is in its home position when it is in its home location with the Up-colored facelet on the Up face.

EXERCISES: (* indicates harder.)

2.2-1   After choosing an Up-face color, how many different choices are possible for the Front-face color?

2.2-2   Which of the following statements are true and which are false?
  a.   When the cube is restored so that each face is a solid color, the colors on opposite faces will always be the same.
  b.   When the positions of the center pieces of two adjacent faces are named then the positional names of all the center pieces are fixed.
  c.   When the positions of any two center pieces are named then the positional names of all the center pieces are fixed.

2.2-3*   How many different orientations of the cube are possible?

3.   NOTATION FOR ABBREVIATIONS

We will abbreviate the six positional names by their initials. To distinguish between the symbols for cubicles and the symbols for cubies, we will use lower case italics for the cubicles and upper case italics for the cubies. Thus the four edge cubicles in the Up layer — shaded in Figure 2-8

EDGE CUBICLES IN THE UP LAYER

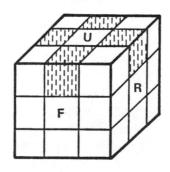

Figure 2-8

— are either denoted *uf, ul, ub,* and *ur,* or denoted *fu, lu, bu,* and *ru.* The four cubies whose home locations are the cubicles *uf, ul, ub,* and *ur* are denoted either by *UF, UL, UB,* and *UR* or by *FU, LU, BU,* and *RU.* The order of the initials of the cubie in the cubicle is used to indicate which facelet is on which face. For example, saying that the *BU* cubie is in the *ur* cubicle means that the Back-colored facelet is on the Up face and the Up-colored facelet is on the Right face. There are twelve edge cubicles and twelve edge cubies denoted as follows:

Cubicles	Cubies
uf or *fu*	*UF* or *FU*
ul or *lu*	*UL* or *LU*
ub or *bu*	*UB* or *BU*
ur or *ru*	*UR* or *RU*
rf or *fr*	*RF* or *FR*
fl or *lf*	*FL* or *LF*
lb or *bl*	*LB* or *BL*
br or *rb*	*BR* or *RB*
df or *fd*	*DF* or *FD*
dl or *ld*	*DL* or *LD*
db or *bd*	*DB* or *BD*
dr or *rd*	*DR* or *RD*

The corner cubicles and cubies are similarly denoted by their facelets. The corner cubicle shaded in Figure 2-9 is the *urf* cubicle. The *urf* cubicle is the home location of the *URF* cubie. The facelets of a corner cubicle or of a corner cubie are always written in a clockwise order as shown in Figure 2-10. The clockwise order in this case is determined by looking along a diagonal line from the outside corner of the cube to the center of the cube. Thus, the three clockwise designations *URF, RFU,* and *FUR* all refer to the same piece whose home location is the *urf* cubicle. That same cubicle is also denoted by *rfu* or *fur.* The counter-clockwise designations, *UFR, FRU,* and *RUF* and *ufr, fru,* and *ruf* are never used. Again the order of the initials is used to show

THE *urf* CUBICLE

DESIGNATING CORNER
FACELETS

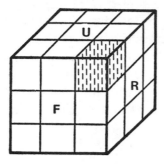

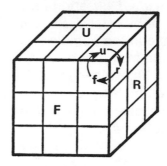

Facelets are designated in
clockwise order.

Figure 2-9   Figure 2-10

the orientation of a cubie in a cubicle. For example, saying
the *BLD* cubie is in the *urf* cubicle means that its Back-colored facelet is on the Up face, its Left-colored facelet is on
the Right face, and its Down-colored facelet is on the Front
face. There are eight corner cubicles and eight corner
cubies denoted as follows:

Cubicles	Cubies
urf, rfu, or *fur*	*URF, RFU,* or *FUR*
ufl, flu, or *luf*	*UFL, FLU,* or *LUF*
ulb, lbu, or *bul*	*ULB, LBU,* or *BUL*
ubr, bru, or *rub*	*UBR, BRU,* or *RUB*
dfr, frd, or *rdf*	*DFR, FRD,* or *RDF*
dlf, lfd, or *fdl*	*DLF, LFD,* or *FDL*
dbl, bld, or *ldb*	*DLB, BLD,* or *LDB*
drb, rbd, or *bdr*	*DRB, RBD,* or *BDR*

We use the order of the facelets when describing the
movement of pieces resulting from a sequence of face
turns. We may say that the piece in the *ufl* cubicle moves to
the *drb* cubicle, and write

$$ufl \rightarrow drb.$$

By this we mean that the facelets of the corner cubie in the *ufl* cubicle move so that the facelet from the Up face moves to the Down face, and the other facelets move with it so that

$$u \rightarrow d$$
$$f \rightarrow r$$
$$l \rightarrow b.$$

When we write

$$rb \rightarrow bu$$

we mean that the edge cubie in the *rb* cubicle moves so that the facelet starting on the Right face moves to the Back face in the *bu* cubicle.

Similarly, we show a clockwise twist of a corner cubie in the *urf* cubicle by writing

$$urf \rightarrow rfu,$$

or the flip of an edge cubie in the *fl* cubicle by writing

$$fl \rightarrow lf.$$

We illustrate the positions of cubies in a cube with a diagram of the type shown in Figure 2-11 which shows all cubies in their home positions. Each initial indicates that the color of the facelet in which it appears is the color of the center cubie with that initial. The initials around the outside of the cube indicate the color of the unseen facelet along the edge adjacent to the initial. No initial is placed in a facelet when its color is unknown. Thus Figure 2-12 illustrates a scrambled cube. As the cube is gradually restored, more and more initials appear.

Once you have established an orientation for your cube, you can move cubies about from cubicle to cubicle by rotating any of the six faces of the cube. These six face rotations are denoted by the initials in block capital letters

$$U, D, F, B, R, \text{ and } L.$$

ALL CUBIES IN THEIR HOME POSITIONS

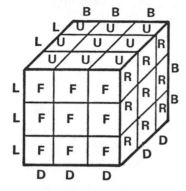

Figure 2-11

A SCRAMBLED CUBE

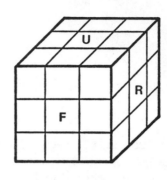

Figure 2-12

A single initial indicates a clockwise quarter turn of the corresponding face. Thus Figure 2-13 shows the cube after the move F has been applied to the cube in its starting state shown in Figure 2-11. The direction of a clockwise quarter turn for any face rotation is defined as shown in Figure 2-14 by viewing that face from that side of the cube. A half turn of

APPLYING THE MOVE F TO A RESTORED CUBE

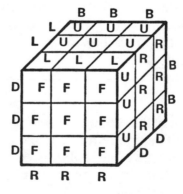

Figure 2-13

CLOCKWISE FACE TURNS

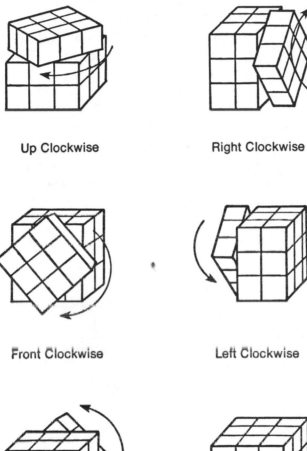

Up Clockwise

Right Clockwise

Front Clockwise

Left Clockwise

Back Clockwise

Down Clockwise

Figure 2-14

any face is two quarter turns of that face. So we use the notation

$$U^2, D^2, F^2, B^2, R^2, \text{ and } L^2$$

to denote half turns of the six faces. The symbol U^2 is pronounced "U squared." Counter-clockwise quarter turns are denoted by

$$U^{-1}, D^{-1}, F^{-1}, B^{-1}, R^{-1}, \text{ and } L^{-1}.$$

The symbol U^{-1} is pronounced "U inverse."

In recording a sequence of moves, we list the moves from left to right. Thus FR means apply F first and then apply R. Figure 2-15 shows the result of applying the sequence FR to a cube in its starting state as shown in Figure 2-11. Any sequence of moves is called a *process*. Sometimes parenthe-

MOVE FR

Figure 2-15

ses are placed around a sequence of moves within a process only to emphasize that those moves are related. Such parentheses can always be ignored without changing the overall process.

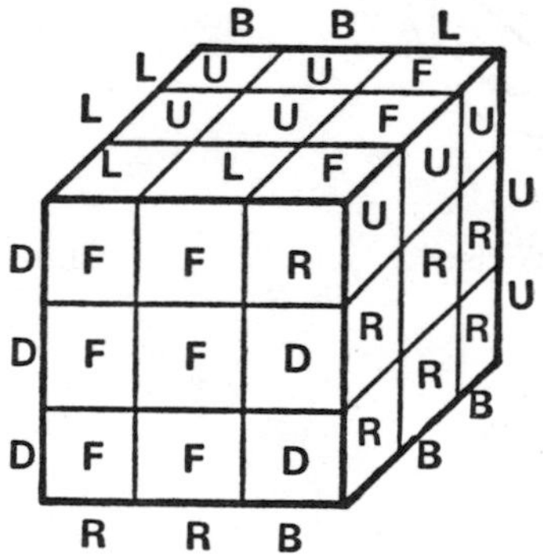

To indicate that a particular process moves cubies from one cubicle to another cubicle, we list the cubicle movement followed by the process. For example, we write

$$uf \rightarrow ru\text{: FR}$$

to indicate that the cubie in the *uf* position is moved to the *ru* position by the process FR.

All of the notation presented so far in this section has referred to the fixed orientation we started with. That is, the locations of the center pieces on each face have remained fixed. However, sometimes it is useful to reorient the cube — perhaps to put the Right face in the front and the Back face on the right or perhaps just to turn the cube over. But, there are several different ways to turn the cube over. We need some notation to describe this more precisely. We use another set of initials to describe the cube reorientations. We use script letters of the initials. The symbol $\mathcal{U}$ — pronounced "script U" — will denote a clockwise quarter turn of the Up face together with the entire cube under it. The symbol $\mathcal{F}$ will denote a clockwise quarter turn of the Front face together with the entire cube behind it. Similarly, $\mathcal{R}$, $\mathcal{L}$, $\mathcal{B}$, and $\mathcal{D}$ will denote clockwise rotations of the entire cube as viewed from the Right, Left, Back, or Down faces respectively. Again we will use $\mathcal{U}^{-1}$, $\mathcal{F}^{-1}$, $\mathcal{R}^{-1}$, $\mathcal{L}^{-1}$, $\mathcal{B}^{-1}$, and $\mathcal{D}^{-1}$ to denote counter-clockwise quarter turns and $\mathcal{U}^2$, $\mathcal{F}^2$, $\mathcal{R}^2$, $\mathcal{L}^2$, $\mathcal{B}^2$, and $\mathcal{D}^2$ to indicate half turns. Notice that

$$\mathcal{R} = \mathcal{U}\mathcal{F}\mathcal{U}^{-1}.$$

This and similar equalities for the other reorientation moves show that the moves $\mathcal{U}$ and $\mathcal{F}$ would have been enough to reorient the cube in any way we wanted. The others still are useful abbreviations.

EXERCISES:

2.3-1   Which edge and corner locations of the cube are not shown by Figure 2-11?

2.3-2   Find a sequence of moves which accomplishes each of the following:

    a)   Moves the piece in the *urf* corner to the *rdf* position.

    b)   Moves the piece in the *rdf* corner to the *fur* corner.

    c)   Moves the piece in the *uf* edge to the *fu* edge position —
that is, flips the *uf* edge piece.

2.3-3   For each sequence of moves used to solve the previous exercise, draw a figure like Figure 2-13 showing the contents of each location on the Up, Right, and Front faces after applying each sequence to a restored cube.

2.3-4   Starting with a restored cube and using the notation presented in this chapter,

    a)   List the Up-layer corner pieces.

    b)   List the Down-layer edge locations.

    c)   List all the pieces in the middle layer between the Up and Down faces whose location can be changed by rotating a face.

    d)   List all the locations to which the *URF* piece can be moved by rotating no more than one face.

    e)   List all the pieces which can be moved to the *uf* location by rotating only one face.

2.3-5   Write expressions using only $\mathcal{U}$ and $\mathcal{F}$ for the three orientation moves $\mathcal{L}$, $\mathcal{B}$, and $\mathcal{D}$.

CHAPTER 3

RESTORING THE CUBE

In this chapter you will learn a method for restoring the cube which has been selected to be as simple and logical as possible. It was chosen from the many known methods because there is no "trick" to be learned and no lengthy sequences of moves to be memorized. Instead, the reason for each move can be understood. You, as student, should understand what you are trying to do at each step and then why the moves which you use will do what you wanted. The restoration method presented here restores one cubie at a time. To restore the entire cube takes a long time. Learning to restore the cube takes much longer, even at best.

Basic Principles. This section presents techniques which form the foundation for understanding much more about the cube than just how to restore it. Basic principles are explained by which you can move unrestored cubies without destroying your previous restoration of the cube. In later chapters of *Handbook of Cubik Math*, more concepts and processes are presented so that you can improve on

21

your restoration method and develop a further understanding of the mathematics embodied in the cube.

All of the principles and techniques which we need to restore the entire cube are exemplified even in the early steps of any restoration process. If you have experimented with your cube, you have undoubtedly applied some of these principles even if you didn't know them. The following is a fundamental principle used throughout the entire restoration procedure.

> *Principle of Inverses.* Any sequence of moves can be reversed and the cubies will be restored to their starting positions. This is called the *inverse* of the process. The inverse of a process Y is denoted by Y^{-1}.

For example, the process $Y = BDRD^{-1}$ can be reversed by the process $Y^{-1} = DR^{-1}D^{-1}B^{-1}$ to restore the cube. Notice that both the order and direction of the moves must be reversed. Looking at one particular piece, we see that the piece in the *db* cubicle is moved to the *fd* cubicle by $BDRD^{-1}$ — see Figure 3-1. So the reverse process $DR^{-1}D^{-1}B^{-1}$ must move the piece in the *fd* cubicle back to the *db* cubicle. Focusing on the individual cubies leads to a corollary of the first principle.

> *Principle of Partial Inverses.* Any cubies moved by a process Y will later be restored to their starting positions by Y^{-1}, the inverse of that process, providing they have not been moved by any other process in the meantime.

To see how we use this principle, consider the following problem. Suppose you have restored three of the Down-face edge cubies. You now want to restore the remaining Down-face edge cubie, the *DF* cubie, to its home position while leaving the previously-restored pieces in their home positions. You find the *DF* cubie is in the *fr* cubicle — that is, it is positioned with its Down-colored facelet on the Front

FOR EXAMPLE, THE PROCESS BDRD⁻¹ MOVES *db* TO *fd*

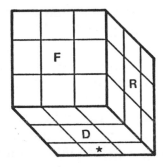

Start: The ★ is in the *db* cubicle.

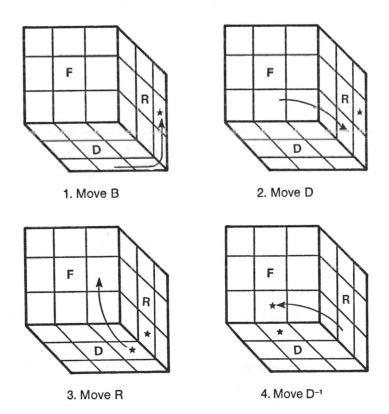

1. Move B

2. Move D

3. Move R

4. Move D⁻¹

Figure 3-1

face and its Front-colored facelet on the Right face. If it were flipped — that is, if it were in the *rf* position — it would be easy. But there is no way by turning only the Front and Up faces to move the Down-colored facelet of the *DF* cubie from the Front face to the Down face. Moving any other faces will move at least one of the previously restored edge cubies out its home position. We need to use the Principle of Partial Inverses.

Following the sequence shown in Figure 3-2, we first turn the Down face so that none of the previously restored cubies is on the Right face. Then turning only the Right face, we place the Down-colored facelet of the DF cubie on the Down face. Now reverse the initial Down-face turn. We have just made use of the Principle of Partial Inverses in the following way. All the pieces which were not moved by the Right face turn — including the three previously restored Down-face edge cubies — are returned to their starting positions. They were moved by the first process D, left in place by the second process R^{-1}, and then returned to their starting positions by D^{-1}, the inverse of the first process.

Conjugate Processes. Any sequence of three processes, X, Y, and Z, in which the last process Z is the inverse of the first process X

$$Z = X^{-1}$$

is called a *conjugate* process. When you write down the three processes in a conjugate, it has the form

$$XYX^{-1}$$

where X^{-1} indicates the inverse of the process X. You should recognize this form in many of the processes used throughout the restoration procedure in this chapter.

Usually when we use a conjugate process, the locations which are moved by the second process Y form a *working space*. The use of a working space is closely linked with the use of conjugates in this restoration procedure. In most

RESTORING *DF* FROM *fr*

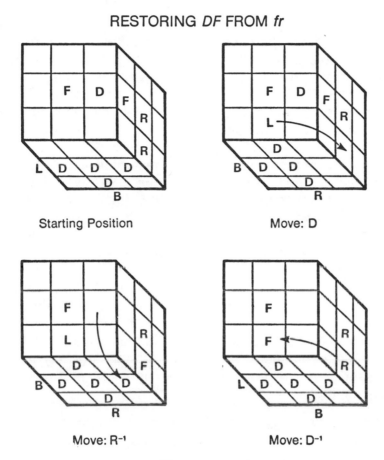

Starting Position   Move: D

Move: R⁻¹   Move: D⁻¹

Figure 3-2

steps of the restoration, conjugates are used extensively. First, a conjugate moves a selected cubie into a working space without destroying the earlier restoration. Second, another conjugate is used to place that cubie into its home position, leaving all the previously restored cubies undisturbed.

When moving the cubie to be restored into the working space, the first process X of the conjugate serves two functions. It keeps the previously restored cubies out of the working space and it moves the selected cubie into the working space. The second process Y then moves the selected cubie so that the third process X⁻¹ leaves the selected cubie in the working space while returning all the previously restored cubies to their home positions.

When restoring the chosen cubie to its home position, the first process X of the conjugate again serves two functions. It moves the previously restored cubies out of the working space. The process X also moves the cubicle to be restored to a cubicle, call it *w*, which is in the working space. The process Y then is used to position the cubie to be restored in the location *w*. Finally the process X⁻¹ not only puts all the previously restored pieces back in their home locations, but also moves the piece to be restored from cubicle *w* to its home position.

You will probably need many examples before you are comfortable with this concept of conjugates. They will be presented as we proceed to explain the restoration procedure. You do not need to understand the concept fully now in order to proceed. But your understanding may be increased if you review this section again several times as you complete later sections in this chapter.

Using Building Blocks. Another technique used throughout this chapter is to build on to earlier solutions. For example, again suppose you have to restore the *DF* cubie after the other three Down-face edge cubies have been restored. But suppose now the *DF* cubie is on the Up face. If the Down-colored facelet is on the Up face, you can rotate the Up face until the *DF* cubie is in the *uf* cubicle and then apply F² to place *DF* in its home position without disturbing the other Down-face edge cubies. But what if the *DF* cubie is in the *fu* cubicle, that is, the Down-colored facelet is on the Front face and the Front-colored facelet is on the Up face? The answer is to rotate the Front face to put the *DF* cubie in the *fr* cubicle. You can then apply the sequence we used earlier, shown in Figure 3-2. Thus we can reduce this problem to one we have previously solved. We will see many more examples of this as we proceed.

The Restoration Sequence. The restoration is done in the following six steps:

Step 1: The Down-face Edge Cubies
Step 2: Three Down-face Corner Cubies
Step 3: Three Middle-layer Edge Cubies
Step 4: The Remaining Five Edge Cubies
Step 5: Placing the Final Corners
Step 6: Untwisting the Final Corners

The darkened cubies in Figure 3-3 indicate the cubies to be restored in each step.

1. THE DOWN-FACE EDGE CUBIES

The first step is to restore all the edge pieces of a single face. Any face will do. So, select one and place the center piece of that color on the Down face. You may want to select a color you don't like, because by selecting it you will get most of it finished and done with early. Or you may prefer to do the most visible color first.

You will probably, with some experimentation, be able to restore the four Down-face edge cubies to their home positions without further instruction. Remember, for these cubies to be in their home positions, not only must the Down-color facelet be on the Down face, but also the other facelet of each restored cubie must be on the side with the center piece of matching color. Try restoring these four pieces for yourself before you read further.

Strategy for Step 1. Restoring the first cubie is particularly easy since there are no previously restored cubies to worry about. But as soon as several cubies have been placed in their home positions, freedom of movement of the cube faces is already restricted. The last cubie to be placed is

SEQUENCE FOR RESTORING CUBIES

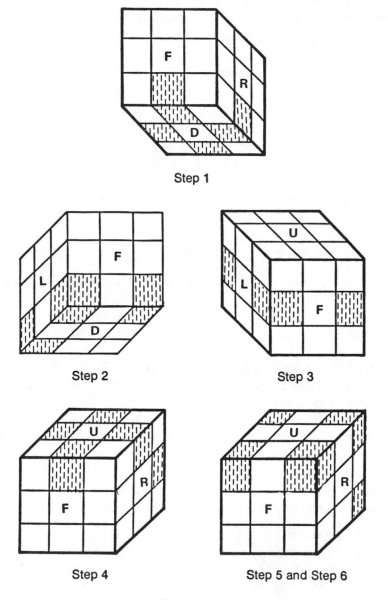

Step 1

Step 2

Step 3

Step 4

Step 5 and Step 6

Figure 3-3

usually the most difficult since it must leave the most other cubies in place. If we can restore any Down-face edge cubie without disturbing the other three when they have been restored, then we can restore the last Down-face edge cubie as well as the earlier Down-face edge cubies.

We will explain how to restore the *DF* edge cubie to the *df* cubicle without disturbing any restored Down-face edges. For the other Down-face edge cubies, we use a simple example of the building block technique discussed at the end of the last section. We reorient the cube, using *U*, until the cubicle to be restored is in the *df* location. Then we can use the same processes to restore the *DF* cubie to the *df* cubicle. Since *U* leaves the Down-face edge cubies on the Down face, the previously restored cubies are not disturbed.

Processes to Restore the *DF* **Cubie.** There are twelve edge cubicles in the cube. The *DF* cubie could be in any of them and in each cubicle the *DF* cubie could be positioned in either of two orientations. Thus there are 24 positions in which we might find the *DF* cubie. Many of these are positions from which it can be placed in the *df* cubicle without moving any previously restored Down-face edge cubies.

Where Previously Restored Cubies Need Not Be Moved. If the Front-colored facelet of the *DF* cubie is on the Front face, then rotating only the Front face will place the *DF* cubie in the *df* cubicle.

$$df \; \rightarrow df$$
$$rf \; \rightarrow df : F$$
$$uf \; \rightarrow df : F^2$$
$$lf \; \rightarrow df : F^{-1}$$

If the Down-colored facelet of *DF* is on the Up face, then rotating only the Up face will place the Front-colored facelet on the Front face. This reduces the problem to the one solved above.

$$ur \rightarrow df : \text{UF}^2$$
$$ub \rightarrow df : \text{U}^2\text{F}^2$$
$$ul \rightarrow df : \text{U}^{-1}\text{F}^2$$

If the Down-colored facelet of the *DF* cubie is on the Down face we can again build on the previous solution.

$$dr \rightarrow df : \text{R}^2\text{UF}^2$$
$$db \rightarrow df : \text{B}^2\text{U}^2\text{F}^2$$
$$dl \rightarrow df : \text{L}^2\text{U}^{-1}\text{F}^2$$

Two other cases can build on the first solution.

$$rd \rightarrow df : \text{RF}$$
$$ld \rightarrow df : \text{L}^{-1}\text{F}^{-1}$$

The other twelve cases all require that at least one of the Down-face edges which may have been restored must be moved in the process of restoring the *DF* cubie to the *df* cubicle.

Using Conjugates to Replace Restored Cubies. We saw in Figure 3-2 of the previous section how the conjugate process DR^{-1}D^{-1} is used to restore the *DF* cubie from the *fr* cubicle leaving the other Down-face edge cubies in place. In a similar manner if the *DF* cubie is in any middle layer cubicle, then the Down layer can be rotated to move the *df* cubicle into a position to receive the *DF* cubie with the Down-colored facelet on the Down face. At the same time the other Down-face edges are moved out of the way of the side rotation needed to place that Down-colored facelet of *DF* onto the Down face. After *DF* is placed in the Down layer, the Down layer is rotated back to its starting position, thus both restoring the *DF* cubie to the *df* cubicle and returning the other Down-face edges to their home positions. The following processes all exemplify this.

$fr \rightarrow df$: DR^{-1}D^{-1}
$br \rightarrow df$: DRD^{-1}
$rb \rightarrow df$: D^{2}B^{-1}D^2
$lb \rightarrow df$: D^{2}BD^2
$bl \rightarrow df$: D^{-1}L^{-1}D
$fl \rightarrow df$: D^{-1}LD

Three other cases can build directly on these conjugates.

$fu \rightarrow df$: FDR^{-1}D^{-1}
$fd \rightarrow df$: FD^{-1}LD
$bd \rightarrow df$: BDRD^{-1}

The last three cases can be built on these also if you want.

$ru \rightarrow df$: UFDR^{-1}D^{-1}
$bu \rightarrow df$: U^{2}FDR^{-1}D^{-1}
$lu \rightarrow df$: U^{-1}FDR^{-1}D^{-1}

However when the *DF* cubie is in either the *ru* or the *lu* position, it can be restored more easily using the conjugates

$ru \rightarrow df$: R^{-1}FR
$lu \rightarrow df$: LF^{-1}L^{-1}.

If the *DF* cubie is in the *bu* cubicle, then building on one of these last two gives

$bu \rightarrow df$: UR^{-1}FR.

This covers the 24 cases for moving the *DF* cubie to the *df* cubicle. We doubt if any of you want to memorize the lists of processes given here. You should examine them sufficiently to be sure you understand the purpose for each move in each process. Once you have that understanding you probably will never have reason to look at this section again.

2. THREE DOWN-FACE CORNER CUBIES

Strategy for Step 2. Having restored the four Down-face edge cubies, the only face that can be rotated without moving these restored pieces is the Up face. In this step and in the following step we will use the Up layer as a working space. We first select a piece to be restored which we call the *selected* cubie, and place that cubie into the working space. We move it from there into its home position. In selecting which should be the next cubie to be restored we look for cubies which are already in that working space. In the process of restoring one of those cubies we will sometimes move another cubie to be restored into the Up-layer working space.

A Well-Prepared Cubie. To move a Down-face corner cubie from the working space to its home position it should be well-prepared. A *well-prepared* corner cubie is one which is oriented in the working space so that its Down-colored facelet is on a side face of the cube — that is, the Down-colored facelet should not be on the Up face. If the Down-colored facelet is on the Up face, then it is more difficult to restore. Placing that corner cubie in its home position without disturbing the previously restored Down-face edge cubies is harder if the cubie has not been well-prepared in the working space.

Well-Preparing a Cubie. When we select a Down-face corner cubie to be restored, it must first be well-prepared. We must move it to the Up layer with its Down-colored facelet on a side face. When we find the selected cubie, one of three things will be true:

1. The cubie is already well-prepared, or
2. The cubie is in the Down layer, or
3. The cubie is on the Up layer with its Down-colored facelet on the Up face.

If it is already well-prepared then nothing more need be done before going on to placing it in its home position. The other two cases are not difficult, but we must be careful not to destroy the previous restoration of Down-face cubies while well-preparing the selected cubie.

Well-Preparing a Selected Cubie from the Down Face.
When we find the cubie we selected to restore on the Down
layer, we first turn the entire cube — that is, reorient the
cube using $\mathcal{U}$— until the selected cubie is in the *dlf* cubicle.
We then move that cubie from the *dlf* cubicle into the Up-
layer working space. One of the two conjugates FU^2F^{-1} or
$L^{-1}U^2L$ will place the selected cubie in a well-prepared orientation in the Up layer. The first of these conjugates moves
the cubie from the *dlf* cubicle to the *rub* position and we write

$$dlf \rightarrow rub : FU^2F^{-1}.$$

This is the process that should be used if, when the selected cubie is in the *dlf* cubicle, its Down-colored facelet is
on the Front face. By using the F move to put the selected
cubie into the Up layer, the Down-colored facelet is kept on
a side face. Figure 3-4 shows this process with an explanation of the purpose of each move.

If, when the selected cubie is in the *dlf* cubicle, the Down-
colored facelet is on the Left face then

$$dlf \rightarrow bru : L^{-1}U^2L$$

should be used to keep the Down-colored facelet on a side
face. If the Down colored facelet started on the Down face
then either process can be used since both will leave the
Down-colored facelet on a side face. Notice how, by using
these conjugates, the previously restored Down-face cubies
are left in their home positions.

Well-Preparing a Selected Cubie from the Up Layer. If we
found the selected cubie on the Up layer with the Down-colored facelet on the Up face then the process is similar. We
can use the same conjugates but first we must orient the
cube so that no previously restored cubies are placed in the
Up layer working space. To do this, use $\mathcal{U}$ repeatedly until
the home position of the selected cubie is placed in the *dlf*
cubicle. In this way we know that the *dlf* cubicle does not
contain a restored cubie. Now rotate the Up face until the

WELL-PREPARING A DOWN-LAYER CUBIE

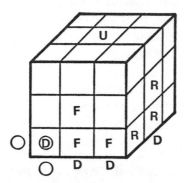

Starting Position
Objective: Move the selected corner cubie — indicated by ○ — to the Up layer with the Down-color facelet on a side face of the cube, and return all previously restored pieces to their home positions.

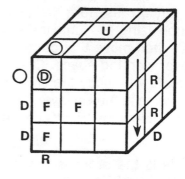

Move F

Move 1: Put the selected cubie into the Up layer with the D-colored facelet on the Front face.

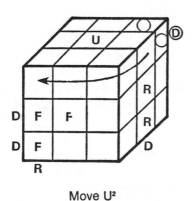

Move U²

Move 2: Place the selected cubie out of the Front layer without moving any previously restored pieces.

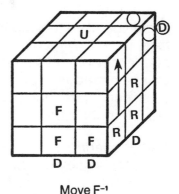

Move F⁻¹

Move 3: Restore the other Down-face cubies moved by Move 1.

Figure 3-4

selected cubie is in the *ufl* cubicle, directly above its home position. Applying either of the two conjugates FU^2F^{-1} or $L^{-1}U^2L$ will moved the Down-colored facelet of the selected cubie to a side face. Again by using these conjugates the previously restored edge and corner cubies are never placed in the Up layer where the work is done, and they are always returned to their home positions at the end of the process.

Placing the Selected Cubie in its Home Position. One of the same two processes used to well-prepare the selected corner in the working space can again be used to place it in its home position. But first the cube must again be oriented — using $\mathcal{U}$ — so that the home position of the cubie to be restored is put in the *dlf* cubicle. Then the Up face must be rotated to place the selected cubie — which should now be called the *DLF* cubie — into the *bru* location. Since the selected cubie has been well-prepared, the Down-colored facelet will be either on the Back face, or on the Right face.

If the Down-colored facelet is on the Back face, we place *DLF* in its home position by using

$$bru \rightarrow dlf: L^{-1}U^2L.$$

This process is shown in Figure 3-5 with an explanation of the reason for each move. By starting with L^{-1}, the *dlf* cubicle is moved to the position in the working space in which it will receive the *DLF* cubie in the correct orientation when U^2 is applied. The inverse process L then moves *DLF* to its home position along with the other previously restored Down-face edge and corner cubies.

If the Down-colored facelet is on the Right face, we place *DLF* in its home position by using the process

$$rub \rightarrow dlf: FU^2F^{-1}.$$

This is the same conjugate process that is shown in Figure 3-4. Again the work is done on the Up layer without destroying the previous restoration of cubies.

It is possible, by reorienting the cube, to restore each of the four Down-face corner cubies in this manner. However,

RESTORING A DOWN-LAYER CORNER CUBIE

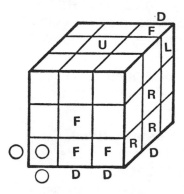

Starting Position
Objective: Move the *DLF* cubie from the *bru* position to its home position in the *dlf* cubicle — indicated by ◯.

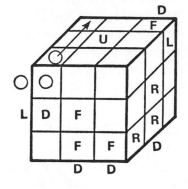

Move L⁻¹

Move 1: Move the *dlf* cubicle to the Up layer with the Down-colored facelet being moved to the Front face.

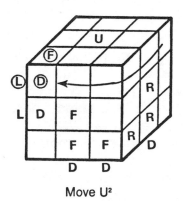

Move U²

Move 2: Place the *DLF* cubie into the Left layer next to the *DL* cubie with its Left-colored facelet on the Left face.

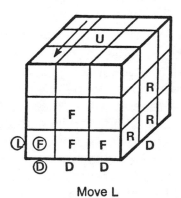

Move L

Move 3: Place the *DLF* cubie into its home position while returning any previously restored cubies to their home positions.

Figure 3-5

only three corners need be restored. The fourth will be used as a working space in later steps and will be called the *working corner*. Its contents will often be changed. Orient the cube so that the *drb* cubicle is the working corner.

3. THREE MIDDLE-LAYER EDGE CUBIES

Having restored the four edge cubies and three of the four corner cubies on the Down face, we now tackle the middle layer between the Up and Down faces. We will restore the three edge cubies whose home positions are directly above the three restored Down-face corner cubies. Again two parts are involved; first, placing the cubie to be restored into the working space — the Up layer — and second, moving that cubie to its home position.

Strategy for Step 3. To avoid moving any previously restored cubies to the working space while moving cubies to and from middle-layer edge cubicles, we always rotate the Down layer to place the unrestored working corner under the middle-layer edge cubicle being moved to the Up layer. A three-move conjugate similar to that used in the previous two steps will then move the selected cubie from the middle layer to the working space or place the selected cubie in its home position from the working space.

Moving a Middle-Layer Cubie to the Working Space. The middle-layer edge cubies are moved to the working space in the same way that corner cubies were moved to the working space in the previous step. If the selected edge cubie is not already in the Up layer, then it must be in the middle layer since the Down-face edges have all been restored. If the selected cubie is in the middle layer then orient the cube so that the selected cubie is in the *fl* location — by repeating *U* as necessary. Then rotate the Down face to put the working corner in the *dlf* cubicle. Either of the two conjugates FU^2F^{-1} or $L^{-1}U^2L$ can be used to move the selected cubie into the Up layer. The Down layer should then be rotated to return the previously restored Down-face cubies to their home positions and the cube should be rotated to return the working corner to the *drb* cubicle.

Placing a Middle-Layer Cubie in its Home Position. Orient the cube — using $\mathcal{U}$ — so that the home position of the selected cubie is in the *fl* cubicle, and rotate the Down layer so that the unrestored working corner is placed in the *dlf* cubicle. Again we may also need to rotate the Up layer so that the selected cubie is not moved out of the working space by the first process of the conjugate. So, to be on the safe side, rotate the Up layer to put the selected cubie into the *ur* cubicle. There are now two possible orientations for the selected cubie in the *ur* cubicle, either

    1. the Front-colored facelet is on the Up face, or
    2. the Left-colored facelet is on the Up face.

Front Color on Up Face. In the first case, we place the *FL* cubie from the *ur* cubicle into its home position *fl* by the process

$$ur \rightarrow fl: \text{L}^{-1}\text{U}^2\text{L}$$

which is again the same process shown earlier in Figure 3-5.

Left Color on Up Face. In the second case, we place the *FL* cubie from the *ru* cubicle into its home position *fl* by the process

$$ru \rightarrow fl: \text{FUF}^{-1}$$

which is only slightly different from the conjugates we have been using. Still the working corner is the only Down-layer cubicle whose contents are changed by this process.

After the selected cubie has been placed in its home position the Down layer should be rotated to return the previously restored Down-layer cubies to their home positions.

This procedure can be repeated to restore all of the middle-layer cubies, but it is not necessary to restore the one over the working corner since that cubicle will be used as a working edge in the next step.

4. THE REMAINING FIVE EDGE CUBIES

After restoring the three middle-layer edge cubies not over the working corner, we are ready to restore the Up-face edges. The unrestored edge of the middle layer will be used as a working edge. But, when we are finished with the Up-face edges we will find that the working edge in the middle layer has also been properly restored.

Now is also the time for another WARNING. At this point you have restored enough of the cube so that carelessness can easily destroy what you have done. It is very frustrating to realize that several moves back you goofed and now must start over. To minimize the possibility of that, whenever we move any of the restored pieces we will continue to operate with conjugate processes consisting of three moves at a time. At the end of each of these three-move sequences, you can check and see that the cubies in the Down and middle layers which started out restored have been returned to their home positions. Between these sequences we will only rotate the Up face. The three-move sequences as in previous steps consist of rotating a side to place the working edge and working corner on the Up face, then rotating the Up face, then rotating the side back to its initial position. We have the cube oriented so that the *drb* cubicle is the working corner. Thus, the three-move conjugates that we will use are $R^{-1}UR$, $R^{-1}U^2R$, $R^{-1}U^{-1}R$, BUB^{-1}, BU^2B^{-1}, and $BU^{-1}B^{-1}$.

Strategy for Step 4. In this step we restore the final four Up-face edge cubies. The processes we use for restoring these four cubies will cause the fifth edge to be placed in its home position automatically. Although the Up-face edge cubies may be restored in any order convenient to you, for ease of presentation we assume that they are to be restored in the order *UL, UF, UB,* and *UR.* The first three are fairly easy. The fourth may require special attention.

Restoring the First Three Up-face Edge Cubies. Again, the restoration of these pieces is in two parts; the first, to move the cubie we select to be restored into the working space — the *rb* cubicle — and the second, to move the selected cubie from the *rb* cubicle to its home position.

Moving an Up-Face Edge Cubie into the Working Edge. To restore the first three Up-face edge cubies, find the cubie to be restored. If it is not already in the working edge cubicle, then it can be moved to the *rb* cubicle by one of the following processes.

$$ul \rightarrow rb : (BUB^{-1})U^{-1}$$
$$uf \rightarrow rb : (BU^2B^{-1})U^2$$
$$ub \rightarrow rb : U^{-1}(BUB^{-1})$$
$$ur \rightarrow rb : (BU^{-1}B^{-1})U$$

without disturbing any previously restored cubies. Parentheses are used only to show the conjugates. Figure 3-6 shows an example of restoring one of the first three Up-face edge cubies with an explanation of the reason for each move. Moves 1 through 3 of that figure show how the conjugate is used to move the selected cubie into the working edge.

Placing the First Three Up-Face Edges in their Home Positions. After the selected cubie has been moved into the working edge *rb*, it can be moved from the *rb* cubicle to its home position by the appropriate choice from the following list:

$$rb \rightarrow uf : U^2(BU^2B^{-1})$$
$$rb \rightarrow fu : U^{-1}(R^{-1}UR)$$
$$rb \rightarrow ul : U(BU^{-1}B^{-1})$$
$$rb \rightarrow lu : U^2(R^{-1}U^2R)$$
$$rb \rightarrow ub : (BU^{-1}B^{-1})U$$
$$rb \rightarrow bu : U(R^{-1}U^{-1}R)$$

RESTORING ONE OF THE FIRST THREE UP-LAYER EDGE CUBIES

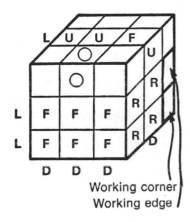

Working corner
Working edge

Starting Position
Objective: Place the *UF* piece in the target location indicated by the ◯ while returning all previously restored pieces to their home positions.

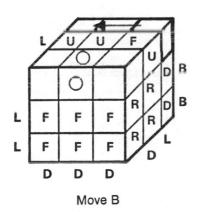

Move B

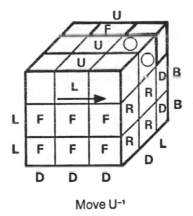

Move U⁻¹

Move 1: Move the working corner and working edge to the Up layer so that the *UF* piece can be put into the working edge cubicle.

Move 2: Place the *UF* piece into position to be moved to the working edge location.

(Continued)

RESTORING ONE OF THE FIRST THREE UP-LAYER EDGE CUBIES (cont.)

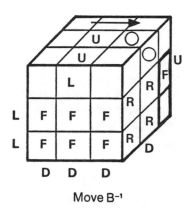

Move B⁻¹

Move 3: Move the piece to be restored, *UF*, into the working edge location, *br*. Now, if it were not already there, it would be necessary to move the target location, ◯, into position to receive the *UF* piece with the U-colored facelet on the Up face.

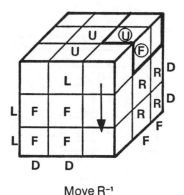

Move R⁻¹

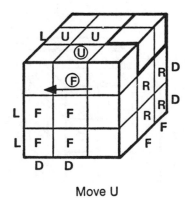

Move U

Move 4: Place the *UF* piece into the ◯ location — that is, one place counter clockwise on the Up face from the *UL* piece — with the U-colored facelet on the Up face.

Move 5: Move *UF* out of the Right layer and into its home position so that the Down face can be restored.

(Continued)

RESTORING ONE OF THE FIRST THREE UP-LAYER EDGE CUBIES (cont.)

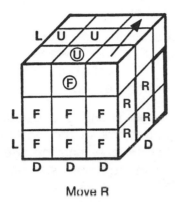

Move 6: Put the previously restored Down-layer and middle-layer pieces back into their home positions while moving the next piece to be restored into the working edge location — and again a new piece into the working corner.

Move R

Figure 3-6

Moves 4 through 6 of Figure 3-6 show how the three-move conjugate is used to place a selected piece in its home position.

Restoring the Fourth Up-Face Edge Cubie. Restoring the last Up-face edge cubie, in this case the *UR* cubie, may require a little special attention. There are three possibilities.

The Lucky Case. The fourth Up-face edge and the cubie in the working edge may already be in their home positions. In this case you can just go on to the next step.

Two Flipped Edges. A second possibility is that the fourth Up-face edge cubie and the cubie in the working edge may both be in their home location but flipped from their home position. In this case we restore this last Up-face edge cubie *UR* in the same way that we restored the first three. We move the *UR* cubie from the *ru* cubicle into the working edge using the process

$$ru \rightarrow br: (BU^{-1}B^{-1})U.$$

Then we place the *UR* cubie into its home position using
the process

$$br \rightarrow ur : (R^{-1}UR)U^{-1}$$

which will also return the *BR* cubie to its home position in
the working edge as well as restoring the Up-face edge
cubie *UB* which was temporarily moved out of place when
the *UR* cubie was moved to the working edge.

The Unlucky Case, A Single Exchange. The third possibility
is that the last Up-face edge cubie *UR* is in the working
edge — which is over the working corner — and the cubie
that belongs in the working edge is in the home position *ur*
of that last Up-face edge cubie. This case is a little different.
You have been unlucky. The properties of the cube are
such that when you are restoring the last face, the last restored *center* cubie must be placed in one of two positions.
These two positions are 180° rotations of each other. By
chance you have chosen to try to restore the cube with the
center cubie turned 90° from one of them. When that occurs, all the Up-face edge cubies which have already been
placed in their home position must be moved a quarter turn
around on the Up face, that is to say that all of their home
positions must be moved one place around relative to the
Up-face center cubie. This will effectively turn the Up-face
center cubie a quarter turn. To do this, we use the one of
the following two processes which moves the Up-colored
facelet of the *UR* cubie to the Up face:

$$rb \longleftrightarrow ur : BUB^{-1}UBUB^{-1}U^2$$
or
$$br \longleftrightarrow ur : U^{-1}R^{-1}U^{-1}RU^{-1}R^{-1}U^{-1}RU^{-1}$$

Notice that the same three-move conjugates are still used in
these processes. The new home positions of the Up-face
edge cubies are moved around on the Up-face center cubie
one position from where they started. Each rotation of the
Up-face within the process moves the new home position of
another Up-face edge cubie in place to receive that cubie

with the next side face rotation. Figure 3-7 shows an example of one of these processes with an explanation of the reason for each move. After the second three-move sequence is completed, all of the edges should be properly oriented in their new home positions including the cubie in the working edge.

RESTORING THE FINAL UP-LAYER EDGE CUBIE

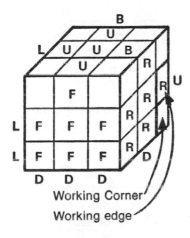

Working Corner
Working edge

Starting Position
Objective: To put all the edge pieces in their home positions, returning all previously restored pieces to their home positions.

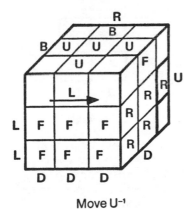

Move U⁻¹

Move 1: The target home position of each Up-face edge piece will be moved one place clockwise. This move prepares the new target home position for the *UR* piece to receive that piece from the working edge location.

(Continued)

RESTORING THE FINAL UP-LAYER EDGE CUBIE (cont.)

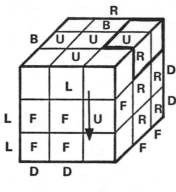

Move R⁻¹

Move 2: Put the *UR* piece into its new home position and move the *UF* piece out of the Up layer.

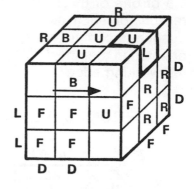

Move U⁻¹

Move 3: Move the *UR* piece out of the Right layer and put the new target home position for the *UF* piece in place to receive that piece.

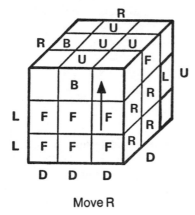

Move R

Move 4: Put the *UF* piece into its new home position and move the *UL* piece out of the Up layer and into the working edge location.

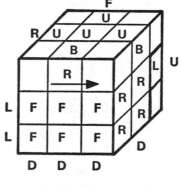

Move U⁻¹

Move 5: Move the *UF* piece out of the Right layer and put the new target home position for the *UL* piece in place to receive that piece.

(Continued)

RESTORING THE FINAL UP-LAYER EDGE CUBIE (cont.)

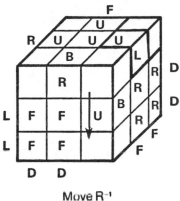

Move R⁻¹

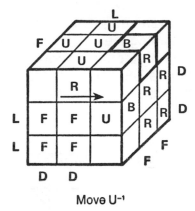

Move U⁻¹

Move 6: Put the *UL* piece into its new home position and move the *UB* piece out of the Up layer.

Move 7: Move the *UL* piece out of the Right layer and put the new target home position for the *UB* piece in place to receive that piece.

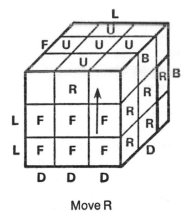

Move R

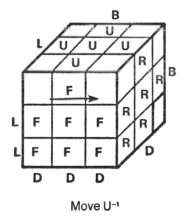

Move U⁻¹

Move 8: Move the *UB* piece into its new home position relative to the other Up-face edges and put the *BR* piece into its home position as well.

Move 9: Place all the Up-face edges into their home positions.

Figure 3-7

5. PLACING THE FINAL CORNERS

All the edge cubies should now have each facelet color matching the color of its adjacent center piece. Thus, each side on the cube should have a cross pattern as shown in Figure 3-8. The corner facelet colors may or may not match the color of the cross, depending on which face you are looking at.

THE EDGE CUBIES MAKE A CROSS ON EVERY FACE

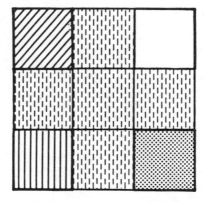

Figure 3-8

The next step is to place the corner pieces in their home locations. In doing this it is not necessary to worry about the orientation of these corners as we will correct the twists of those corners in a later and final step. However, you must be very careful not to disrupt the edge pieces which now have been put in place. Of course the rotation of any face changes the location of both corner and edge pieces. But it is possible to combine a sequence of conjugate processes so that each previously restored cubie which is moved is later put back where it came from. At the end of the total process only corner pieces will have been moved.

Strategy for Step 5. We will move one corner cubie, the *selected corner*, on the Up face to another corner cubicle, called the *target corner*, on the Up face without disturbing the edge pieces. This total process of Step 5 is done again in two parts. First the selected corner cubie is exchanged with the cubie in the working corner *drb*. Then in the second part the selected corner cubie is moved from the working corner into the target corner.

This procedure makes use again of the Principle of Partial Inverses. The process used to place the selected corner into the working corner moves several other cubies in the middle and Down layers. Then the inverse of that process is used in moving the selected corner cubie from the working corner to the target corner. This second part thus restores those middle- and Down-layer cubies to their original home positions. The only difference between the second part and the inverse of the first part is a rotation of the Up layer.

You may be tempted at some time to select a piece that is already in the working corner and place it directly into the target corner, hoping to avoid the first part of this step — moving it into the working corner. However, if you do try this you cannot stop at this point. You must still apply the Principle of Partial Inverses to restore the middle layer and Down layer of the cube. To do this you must exchange another Up layer corner with the working corner. This second part of the process provides the inverse for the middle and Down layers.

A Process for Relocating an Up-Layer Corner Cubie. Orient the cube — using $\mathcal{U}$ — so that the working corner is in the *drb* cubicle. Then rotate the Up layer so that the selected corner cubie is in the *ufl* cubicle. We can use the process

$$ufl \longleftrightarrow rbd : F^{-1}D^2F$$

to interchange the selected corner and the working corner. Again rotating the Up layer we can return the previously restored Up layer cubies to their home position. The only cubies to be moved other than the selected corner and working corner are cubies in the middle and Down layers. We now rotate the Up layer so that the target corner is placed in the *ufl* cubicle. Again we use the process

$$ufl \longleftrightarrow rbd : F^{-1}D^2F$$

this time to interchange the target corner cubie and the selected corner cubie which is in the working corner cubicle. The Up layer is again rotated to restore the Up-face edge cubies. The final result is that the only cubies which are moved are the three corner cubies, the selected corner to the target corner, the target corner to the working corner, and the working corner to the selected corner. Figure 3-9 presents an example of this procedure with an explanation of the purpose of each move.

RELOCATING AN UP-LAYER CORNER CUBIE

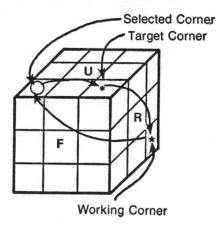

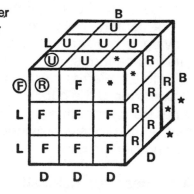

Working Corner

Starting Position

The piece in *ufl* moves to location *urf*. The piece in *urf* moves to location *rbd*. The piece in *rbd* moves to *ufl*.

Objective: Place the selected corner, indicated by ○, into the target corner location, indicated by •, while returning all previously restored pieces to their home positions.

(Continued)

RELOCATING AN UP-LAYER CORNER CUBIE (cont.)

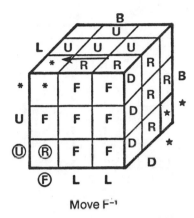

Move F⁻¹

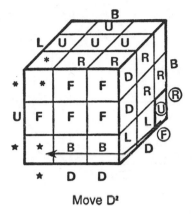

Move D²

Move 1: Place the selected piece *URF* into the Down layer. (A rotation of the Left face could equally well have been used here.)

Move 2: Place the selected corner, *URF*, temporarily into the working corner location. The piece from the working corner takes the place of the selected corner on the Front layer.

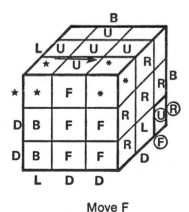

Move F

Move 3: Restore the Up layer with the piece from the working corner in place of the selected corner. (If the left face had been rotated in Move 1 then it also would have to be used here to restore the Up layer. The orientation of the piece from the working corner would have been different.)

(Continued)

RELOCATING AN UP-LAYER CORNER CUBIE (cont.)

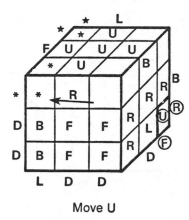

Move U

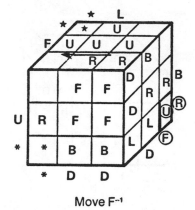

Move F⁻¹

Move 4: Move the target corner piece, *, into the place of the selected corner. (Any Up-face corner could have been chosen as the target corner.)

Move 5: Reversing Move 3, put the piece, *, from the target corner, into the Down layer. (Again if the Left face was used in Move 3, it must be used here.)

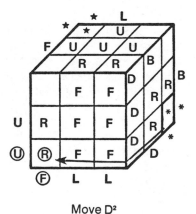

Move D²

Move 6: Reversing Move 2, the target corner piece, *, and the selected corner piece, *URF*, are exchanged on the Down layer. The Down layer is partially restored.

(Continued)

RELOCATING AN UP-LAYER CORNER CUBIE (cont.)

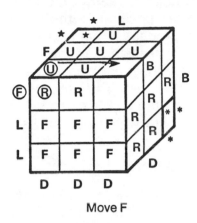

Move F

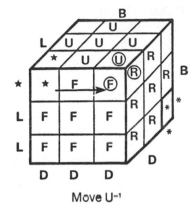

Move U⁻¹

Move 7: Put the selected corner piece, *URF*, into the location vacated by the target corner piece, ∗. This reverses Move 1, thus completing the restoration of the Down layer and the middle layer. (Again, the Left face must be used if it was used in Move 1.)

Move 8: Move the selected corner piece, *URF*, to the target corner location, in the process, returning the Up-face edge pieces to their home positions. That the *URF* piece is oriented correctly in its home position is fortuitous. Whatever facelet of *URF* was on the Up face in the selected corner will still be on the Up face when it has been moved to the target corner location.

Figure 3-9

To see that this is another example of the Principle of Partial Inverses, observe that the inverse of the process $F^{-1}D^2F$ — that is the reverse sequence of moves in the reverse reverse direction — is again the process $F^{-1}D^2F$. Therefore, as we saw with earlier conjugate processes, any cubies which are not moved to the Up layer by $F^{-1}D^2F$ when using a process in the first part will be returned to their starting positions by $F^{-1}D^2F$ when you use that process for the second time.

An Alternate Process. You may have noticed that this conjugate $F^{-1}D^2F$ is similar to the conjugates used in Step 2 to place corner cubies in their home position in the Down layer. The difference is that the process is being used to bring corner cubies into the Up layer instead of the Down layer without disturbing the other Up-layer cubies. It is as if the cube had been turned upside-down. The process D^2 plays the same role as the process U^2 did in Step 2. Just as in Step 2, there is another conjugate which can be used here in Step 5. The process

$$ufl \longleftrightarrow bdr : LD^2L^{-1}$$

can be used in place of $F^{-1}D^2F$ to interchange the selected corner cubie and the cubie in the working corner. But if LD^2L^{-1} is used in the first part of Step 5, then LD^2L^{-1} must also be used in the second part to bring the selected corner cubie back into the target corner cubicle. Since the inverse of LD^2L^{-1} is LD^2L^{-1}, again the Principle of Partial inverses guarantees that the previously restored middle-layer and Down-layer cubies will be returned to their home positions.

Why should you choose one of these processes over the other? The difference is in the orientation of the working corner cubie when it moves to the selected corner cubicle and of the target corner cubie when it moves to the working corner cubicle. By choosing between the two conjugate processes $F^{-1}D^2F$ and LD^2L^{-1} the number of corners that need to be untwisted in the next step can be reduced.

The Next Corner. Continue to move corners of the Up layer into their correct locations until all the corners have been properly located, although not necessarily with proper orientation. If you are careful, at most three applications of this corner process should be required before all corner cubies will be in their home locations. However, anywhere from zero to five of them may need to be twisted. The three corner cubies which were properly placed and oriented in Step 2 on the Down face should still be in their home positions.

6. UNTWISTING THE FINAL CORNERS

All edge cubies should now be in their home positions and all corner cubies should be in their home locations. The final step will reorient — that is, untwist — any corner cubies which are not already in their home positions. The directions — clockwise or counter-clockwise — for corner twists are defined by viewing the cube from the outside along a diagonal through the corner to the center of the cube. For example, if the *URF* cubie is in the *rfu* cubicle, as shown in the starting position of Figure 3-11 with the Up-colored facelet on the Right face, then it is twisted clockwise from its home position. It needs to be twisted counter-clockwise in order to restore it to its home position. The properties of the cube are such that when one corner is twisted one way another corner also must be twisted to compensate. One corner twisted in the opposite direction will compensate. Also, two other corners twisted in the same direction can compensate since two twists in the same direction are the same as one twist in the other. The result must be that if all the corner twists were applied to a single corner, that corner would end up looking as if it had not been twisted at all. This is a property of the cube.

Strategy for Step 6. We will show how to twist one corner of the Up face and then twist another corner of the Up face in the opposite direction. The first twist again messes up cubies in the Down and middle layers. But, the second twist — in the opposite direction — reverses the process and restores those middle-layer and Down-layer cubies to their positions before the two twists were made. The only Up-layer cubie moved by a twist is the corner being twisted. Therefore, turning the Up face before reversing the twist only changes the corner cubie to be untwisted.

It is important to remember that the second twist must be the reverse of the first in order to restore the middle and Down layers of the cube even though the second piece might need to be twisted in the same direction as the first.

Thus, the Up-face corner cubies must be untwisted in pairs. One corner must be twisted clockwise and the other must be twisted counter-clockwise.

Processes to Twist and Untwist Corner Cubies. Orient the cube — using $\mathcal{J}$ or $\mathcal{R}$ if necessary — so that two corners to be twisted are in the Up layer. Decide which one is to be twisted clockwise and which is to be twisted counter-clockwise. Rotate the cube — using $\mathcal{U}$ — so that the corner to be twisted clockwise is in the *ufl* cubicle. We use the process

$$ufl \rightarrow flu : (LD^2L^{-1})(F^{-1}D^2F)$$

to twist the corner clockwise. Now rotate the Up layer so that the corner to be twisted counter-clockwise is in the *ufl* cubicle. We use the process

$$ufl \rightarrow luf : (F^{-1}D^2F)(LD^2L^{-1})$$

to twist the cubie in the *ufl* cubicle counter-clockwise. The counter-clockwise twisting process is the inverse of the clockwise twisting process. Rotating only the Up layer between twists left the middle- and Down-layer cubies alone. Thus, the Principle of Partial Inverses guarantees that the middle- and Down-layer cubies that were moved by the first twist are returned to their home positions by the second twist in the opposite direction. An example illustrating the untwisting of two corners is given in Figure 3-11 with an explanation of the purpose of each move.

Let us emphasize again that even when you find that your cube has three corners that need to be twisted in the same direction, you must start by twisting two of them in opposite directions. You must apply the Principle of Partial Inverses to avoid messing up the middle and Down layers and ruining all the work that you have done to this point. In later chapters of the *Handbook of Cubik Math* techniques are taught for twisting three corners in the same direction, but for now you should be content to twist and untwist pairs.

UNTWISTING TWO UP-LAYER CORNER CUBIES

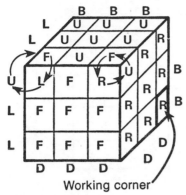

Working corner

Starting Position

Objective: Twist one Up-face corner piece clockwise and twist another Up-face corner piece counter-clockwise, leaving all other pieces of the cube in starting positions at the end. The arrows indicate the required corner twists.

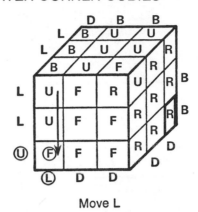

Move L

Move 1: Move the first corner to be twisted, *UFL*, indicated by ○, to the Down layer by rotating one of the side faces. Choose the side face which contains the Up-colored facelet of the piece to be twisted.

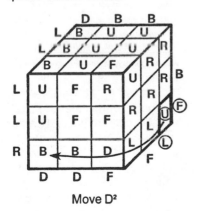

Move D²

Move 2: Put the piece to be twisted, *UFL*, into the working corner, temporarily out of the way.

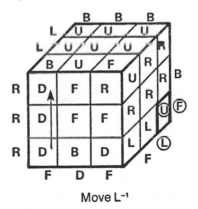

Move L⁻¹

Move 3: Restore the Up face to its starting position except for the location of the piece to be twisted.

(Continued)

UNTWISTING TWO UP-LAYER CORNER CUBIES (cont.)

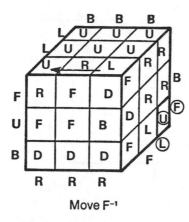

Move F⁻¹

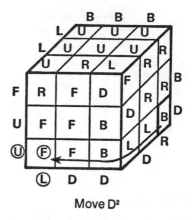

Move D²

Move 4: Using the other side layer which contains the starting location of the piece to be twisted, move that location back to the Down layer in preparation for retrieving the piece in a different orientation.

Move 5: Again exchange the piece from the working corner and the piece to be twisted. This puts the *UFL* corner piece next to the *UF* edge with the correct orientation.

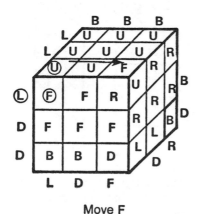

Move F

Move 6: Put the *UFL* piece into its home position untwisted. The pieces in the Up layer are now back to their starting positions except that the first corner has been twisted. We must now restore the Down layer and the middle layer by untwisting another Up face corner with the reverse of the twisting processes.

(Continued)

UNTWISTING TWO UP-LAYER CORNER CUBIES (cont.)

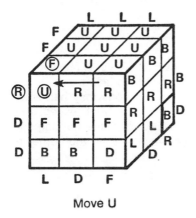

Move U

Move 7: Put the second corner to be untwisted, *URF*, now indicated by ◯, in the home location of the first corner that was twisted. Rotate the Up face here so as not to change the Down layer and the middle layer. Otherwise they would not be restored by reversing the twisting process.

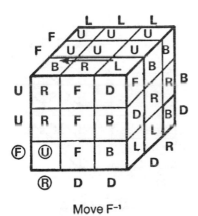

Move F⁻¹

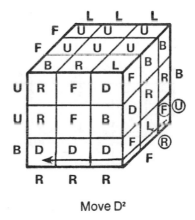

Move D²

Move 8: (Reverse Move 6) Put the second corner to be twisted, *URF*, into the Down layer. If this twist is to restore the corner to its home position then the Up-colored facelet of the corner will be on the face that is rotated by this move.

Move 9: (Reverse Move 5) Put the second piece to be twisted into the "working corner" temporarily out of the way.

(Continued)

UNTWISTING TWO UP-LAYER CORNER CUBIES (cont.)

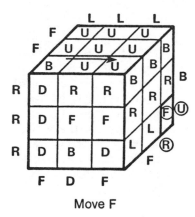

Move F

Move 10: (Reverse Move 4) Restore the Up-face pieces to the same positions they had when starting this twist, except that the piece from the working corner has replaced the corner to be twisted.

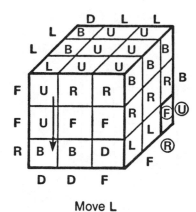

Move L

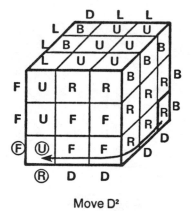

Move D²

Move 11: (Reverse Move 3) Move the target location of the corner being twisted, containing the piece from the working corner, to the Down layer using the other side so as to change the orientation of the corner to be twisted when it is returned.

Move 12: (Reverse Move 2) Place the piece from the working corner back in its home position and the corner to be twisted, *URF*, next to the *UF* edge with the correct orientation.

(Continued)

UNTWISTING TWO UP-LAYER CORNER CUBIES (cont.)

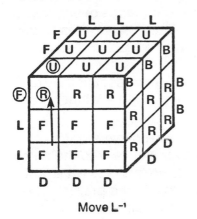

Move L⁻¹   Move U⁻¹

Move 13: (Reverse Move 1) Restore the Down layer and middle layer pieces to their home
positions. Place the Up-face
pieces in their starting positions relative to each other except for the twists of the two
corners.

Move 14: (Reverse Move 7)
Place the Up-face pieces into
their home positions.

Figure 3-11

To untwist the cubie in the working corner, the *rbd* cubicle, reorient the cube to place that corner on a new Up face
together with some other corner that needs to be twisted.

When the next to last corner has been twisted to its home
position, then the last piece will automatically be correctly
oriented in its home position. When that is done, it is time to
celebrate your conquest. You can now restore any scrambled cube!

7. CUBIK GAMES

In place of exercises for this chapter, we suggest you try
out some of the following Cubik Games.

I: Solve It — One or more players.

Have a non-player apply the same sequence of either 4, 5, or 6 moves to each player's cube. Each player tries to discover the unknown sequence of that length to restore the cube. The first to discover it wins. Note: If only 3 moves are used it is too easy. If 7 or more moves are used it may take several hours of concentrated effort to find a winner.

II: Scrambled Cubing — Two or more players.

Have a non-player put each player's cube into the same scrambled starting state. One way to do this is to write down a sequence of face rotations unknown to the players. Then taking each player's cube in the restored state, apply the written sequence of face rotations to each of the cubes. Each player is then to restore his own cube in as few moves as possible. Each rotation of a face — clockwise, counterclockwise, or halfway around — counts as a single move. The player who restores his cube in the fewest moves wins.

III: Contest Cubing — Two players exactly.

Each player scrambles the other player's cube and returns it. Each player is then to restore his cube in as few moves as possible. Again, each rotation of a face — clockwise, counterclockwise or halfway around — counts as a single move. The player who restores his cube in the fewest moves wins.

Variations. The following variations can be used with either *Scrambled Cubing* or *Contest Cubing* to include speed of restoration.

A. Time the restoration process and if the total restoration time is over ten minutes, then add a penalty of one move for each ten seconds over ten minutes.

B. Let the player who restores his cube in the fastest time be the winner.

C. Reduce the score of the first, second, and third fastest finishers by 20 moves, 10 moves, and 5 moves respectively.

IV: Edging — Two to twelve players.

Scramble a cube. Each player chooses an edge of the cube. The object of the game for each player is to be the first one to have the three pieces on his chosen edge of the cube restored — even if done accidentally by another player. Each player sequentially is allowed three moves of the cube. The first move may not rotate the same face as the last move of the preceding player. Different skill levels can be accommodated by allowing the players, a different number of moves at each turn.

V: Face-to-Face — Two players.

Scramble a cube. Player I is to restore the Right face of the cube. Player II is to restore the Left face. Each player in turn is allowed up to eight moves of the cube except that the first move of each player must rotate a different face from the last face rotated by the opponent. The first player to have all the pieces on his face in their home positions wins the game.

Variations. These variations can be adapted to either Game IV or Game V.

A. Scramble the cube turning only the Front and Right faces. Player I is to restore the Right face and Player II is to restore the Front face. Again each player in turn is allowed eight moves except that only rotations of the Front and Right faces are allowed.

B. A time limit may be placed on each player's turn. When the time has expired the remainder of that player's turn is skipped.

C. The number of moves allowed for each player's turn may be varied.

D. Let Player I have one move, then let each player in turn have two moves, then let each player in turn have three moves. Continue to let each player in turn have one move more than on his previous turn until one player wins.

CHAPTER 4

THE WHAT, WHY, AND HOW OF CUBE MOVEMENTS

Now that you can restore the cube, do you think you would win a $100 prize by restoring it in less than five minutes? When a local department store offered that prize to promote sales of the cube, very few people could do it even though they knew a restoration method. On the other hand, many school students have average times of less than two minutes and there are several with average times around 45 seconds per cube to restore a large number of randomly scrambled cubes! But pure speed does not usually indicate an understanding of the mathematics of the cube but rather dexterity, practice, and a smooth-sanded well-greased cube.

A more interesting challenge for students of cubik math is to minimize the number of moves required to restore the cube from any particular configuration. How many moves does one need to restore a randomly scrambled cube? What configurations of the cube are hardest to restore? Nobody knows the best answers to these questions. Throughout this chapter, we will develop new processes requiring fewer moves which will improve your cube restoration ability. The concepts which are used to develop these processes are concepts such as permutations, identities, inverses, commutators, and conjugates which occur throughout mathematics and science. After you understand how these relate to moving pieces about on the cube, you may be able to apply these same concepts in other fields, particularly mathematics, physics, computer science, and engineering.

1. PROCESSES AND PERMUTATIONS

Any *process* — that is, sequence of face rotations — results in a rearrangement of the $54 = 9 \times 6$ facelets of the cubies in the cube. Actually only 48 facelets need to be considered since we have already observed that no center piece ever changes Its location. Such a rearrangement of facelets is called a permutation. Any rearrangement of a finite set of objects is called a *permutation* of those objects.

To describe a permutation of the cubies of the cube, we construct a list showing where the piece in each cubicle is moved to and indicating the new position of each facelet of each cubie. For example, it the piece in the Up-Front-Left corner is moved to the Right-Back-Down corner location with the Up face going to the Right-face position we would write

$$ufl \rightarrow rbd.$$

The permutation caused by a single clockwise quarter turn of the Up face would result in moving the pieces from the locations listed in the left column below to new positions listed in the right column as indicated by the arrows.

$$uf \;\; \rightarrow ul$$
$$ufl \;\; \rightarrow ulb$$

$$ul \; \rightarrow ub$$
$$ulb \rightarrow ubr$$
$$ub \; \rightarrow ur$$
$$ubr \rightarrow urf$$
$$ur \; \rightarrow uf$$
$$urf \rightarrow ufl$$

Twist the Up face of your restored cube to check this out.

If we apply U to a restored cube, then the *UF* cubie is moved to the position *ul* where the *UL* cubie was, etc. However, if we apply U to a scrambled cube, the effect is to move whatever piece is in the *uf* cubicle to the *ul* cubicle, etc. We shall refer to the *uf* cubicle — or any other cubicle — as if it were an object and say, for example, that U moves *uf* to *ul*.

If a particular cubicle is left unmoved by a permutation then it is conventional to leave that location out of the list. For example,

$$dfl \rightarrow dfl$$

is not included in the list above. However, if a piece in a location is flipped or twisted by a permutation, then it must be included in the list. Thus, the permutation of the process UR — that is, a clockwise quarter turn of the Up face followed by a clockwise quarter turn of the Right face — is described by the following list:

$$ur \; \rightarrow uf$$
$$urf \rightarrow ufl$$
$$uf \; \rightarrow ul$$
$$ufl \rightarrow ulb$$
$$ul \; \rightarrow ub$$
$$ulb \rightarrow bdr$$
$$ub \rightarrow br$$
$$ubr \rightarrow bru$$
$$rb \; \rightarrow rd$$
$$drb \rightarrow frd$$
$$rd \; \rightarrow rf$$

$$frd \longrightarrow urf$$
$$rf \longrightarrow ru$$

Again, check this out on your cube. The eighth entry indicates that the cubie in the *ubr* cubicle stays in the same location, but is twisted by 120° clockwise — viewed from outside the cube along the diagonal through the corner and the center of the cube. We observe that the permutation of the process UR can be obtained from the permutation of U followed by the permutation of R as shown by the following diagram. This is called the product of U and R.

Apply U Apply R

$$ur \longrightarrow uf$$
$$urf \longrightarrow ufl$$
$$uf \longrightarrow ul$$
$$ufl \longrightarrow ulb$$
$$ul \longrightarrow ub$$
$$ulb \longrightarrow ubr \longrightarrow bdr$$
$$ub \longrightarrow ur \longrightarrow br$$
$$ubr \longrightarrow urf \longrightarrow bru$$
$$rb \longrightarrow rd$$
$$drb \longrightarrow frd$$
$$rd \longrightarrow rf$$
$$frd \longrightarrow urf$$
$$rf \longrightarrow ru$$

These lists can describe any permutation, not just permutations of the cube. Permutations may also be written in a more condensed manner, which turns out to be very informative. To obtain this condensed form, we consider a cubie in some cubicle *x* and the sequence of cubicles through which it passes as the same permutation is repeated. If in the permutation, *x* is moved to *y*, and *y* is moved to *z*, and *z* is moved back to *x*, then we write

$$(x, y, z)$$

where the closing of the parenthesis indicates that the cubie in the last cubicle is moved back to the first cubicle. This is called a *cycle* or, more specifically, a *3-cycle*, corresponding to the permutation indicated by this diagram.

We could equally well have started with *y* or *z*, so we see that the cycles *(x, y, z)*, *(y, z, x)*, and *(z, x, y)* describe the same permutation.

For example, the permutation of the process U^2R^2 is described by the following list:

$$
\begin{aligned}
uf &\rightarrow ub \\
ufl &\rightarrow dfr \\
ul &\rightarrow dr \\
ulb &\rightarrow drb \\
ub &\rightarrow uf \\
ubr &\rightarrow ufl \\
ur &\rightarrow ul \\
urf &\rightarrow ulb \\
rf &\rightarrow rb \\
dfr &\rightarrow ubr \\
dr &\rightarrow ur \\
drb &\rightarrow urf \\
rb &\rightarrow rf
\end{aligned}
$$

If we let *x* be the cubicle *ufl*, then we see that *x* moves to *y= dfr*, *y* moves to *z= ubr*, and *z* moves to *x= ufl*. These moves form the 3-cycle *(ufl, dfr, ubr)*.

At this point you may wonder if, whenever we write a sequence of cubicle moves determined by a permutation, it always forms a cycle, that is, it always eventually returns to the first cubicle. The following argument shows that the answer is yes.

Since there are only a finite number of cubicles, the sequence must eventually repeat some cubicle. Since a permutation carries just one cubie to each cubicle and every cubicle in the sequence, except the first, already has a cubie carried to it, then the only cubicle which can be repeated is the first cubicle in the sequence. For example, we cannot have

$$x \rightarrow y \rightarrow z \rightarrow y$$

because this has both x and z carried to y. The last cubicle in the sequence must cycle back to its first cubicle. Thus, the sequence must be cyclic — that is x must come back to its starting position so that the sequence will start over again. The sequence of cubicles through which x passes when the permutation is repeated is called the *cycle determined by* x. For example, consider the process UR whose permutation in list form is given above. The cycle determined by the edge cubicle *ur* is

(*ur, uf, ul, ub, br, dr, fr*).

The *cycle representation* of a permutation is obtained as follows. We start by choosing any cubicle x and find the cycle it determines. We then take any cubicle not included in the previous cycle and find the cycle it determines. This cycle cannot involve any cubicles in the previous cycle, or else some two cubies would be carried to the same cubicle. Therefore we say these cycles are *disjoint*. Continuing until no cubicles are left, we have decomposed our permutation into disjoint cycles. The order in which we write disjoint cycles does not matter. For example, the permutation of an Up face quarter turn U is described by the two cycles

(*ufl, ulb, ubr, urf*)

showing the rearrangement of the corner pieces illustrated in Figure 4-1, and

(uf, ul, ub, ur)

showing the rearrangement of the edge pieces also illustrated in Figure 4-1. It does not matter which cycle we write first.

For the process U^2R^2 whose permutation is listed above we saw that the cycle determined by the cubicle *ufl* was

(ufl, dfr, ubr).

Continuing, we see that the *ulb* cubicle determines the cycle

(ulb, drb, urf)

and no other corner cubicles are moved by U^2R^2. Decomposing a permutation into all of its disjoint cycles gives the cycle representation of that permutation. Thus, the permutation of the process U^2R^2 has the following cycle representation:

(ufl, dfr, ubr)
(ulb, drb, urf)
(ul, dr, ur)
(uf, ub)
(rf, rb).

We use a shortened notation for twists and flips in the cycle representation of permutations. For example, in the process UR, where we have

ubr → bru,

we could treat each orientation of that corner piece separately and write *(ubr, bru, rub)*. That is cumbersome and instead, we write

(ubr)$_+$

AN UP-FACE QUARTER TURN

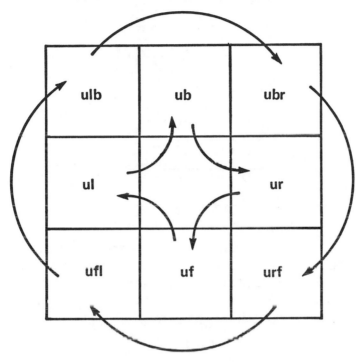

Figure 4-1

to indicate that the last piece returns to the place of the first piece rotated 120° clockwise — that is, the Up face to the Back, the Back face to the Right, and the Right face to the Up, see Figure 4-2. The rest of the permutation of the UR process is described in cycle representation by

(ur, uf, ul, ub, br, dr, fr)

and

*(urf, ufl, ulb, bdr, dfr)_*

where the minus indicates that *dfr* goes to the *urf* place rotated counter-clockwise by 120° — that is, to the *fur* place.

THE PROCESS UR APPLIED TO A RESTORED CUBE

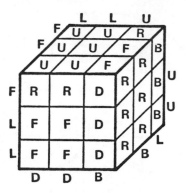

Figure 4-2

Of course, cycles of edges and of corners are disjoint since one type can't move to the place of the other type. A plus sign at the end of a cycle of edges indicates that the cubie in the last edge position moves to the first edge cubicle but in the flipped orientation. For example,

$$(uf, rf)_+$$

indicates that *rf* goes to *fu*.

We will usually use the cycle representation to describe a permutation although occasionally we will find it useful to use the list form as well.

EXERCISES:

4.1-1   Find a sequence of moves which moves *ufl* to *rbd*. Describe the rest of the permutation produced by that sequence of moves.

4.1-2   Write the permutation of the process UR⁻¹ in both the list form and in the cycle representation.

4.1-3   Write, in the cycle representation, the permutation of each of the following processes:
   a)   URU⁻¹R⁻¹
   b)   LD²L⁻¹U²LD²L⁻¹U²

   c) $LR^{-1}F^2L^{-1}RU^2$
   d) $(UR^{-1}U^{-1}R)^2 = UR^{-1}U^{-1}RUR^{-1}U^{-1}R$

4.1-4   Find a process which produces each of the following permutations:

   a)   *(fur, luf, bul, rub) (ur, uf, ul, ub)*
   b)   *(ulb, flu, drb)*
   c)   *(ufl)$_+$ (drb)$_-$*

2. EQUIVALENT PROCESSES

Every process defines a unique permutation which results from applying that process, but the converse is not true. It is *not* true that each permutation comes from a unique process. Many different processes can result in the same permutation. We define two processes to be *equivalent* if and only if they result in the same permutation. Thus, the processes U^{-1} and U^3 are equivalent. Also, FB and BF are equivalent processes. These are considered trivial equivalences. A computer search was conducted to find the shortest pair of non-trivially equivalent processes. The smallest pair of non-trivially equivalent processes required at least eight moves for the two processes combined. One example that was found was that the processes $F^2B^2L^2R^2$ and $R^2L^2B^2F^2$ are equivalent. They each produced the permutation

(ufl, ubr)	*(uf, df)*
(urf, ulb)	*(ul, dl)*
(dlf, drb)	*(ub, db)*
(drf, dbl)	*(ur, dr).*

This happens to be quite a pretty pattern on the cube. Notice that FBLR and RLBF are not equivalent processes since FBLR produces the permutation

(urf)$_+$ (ulb)$_+$ (drb)$_+$ (dlf)$_+$
(ufl, rub)$_+$ (dfr, ldb)$_+$
(uf, ru, rb, ub, lu, lf)
(df, ld, lb, db, rd, rf)

while RLBF produces the permutation

$$(ufl)_+ (ubr)_+ (dfr)_+ (dbl)_+$$
$$(urf, bul)_+ (dlf, bdr)_+$$
$$(uf, rf, ru, ub, lb, lu)$$
$$(df, lf, ld, db, rb, rd).$$

Prove this to yourself by trying it on your cube.

EXERCISES:

4.2-1   Show that the following pairs of processes are equivalent:
   a) $F^2B^2L^2$ and $R^2L^2B^2F^2R^2$
   b) $(F^2R^2)^2 = F^2R^2F^2R^2$ and $(R^2F^2)^4 = R^2F^2R^2F^2R^2F^2R^2F^2$
   c) $F^{-1}D^2FU^2F^{-1}D^2FU^2$ and $U^2RDR^{-1}U^2RD^{-1}R^{-1}$
   d) $FR^{-1}F^{-1}R$ and $U^{-1}RUR^{-1}F^{-1}UFU^{-1}$
4.2-2   Show that FR is *not* equivalent to RF.
4.2-3   In mathematics, an equivalence relation should have three properties:
   i) Any element, X, must be equivalent to itself — Reflexive.
   ii) If X is equivalent to Y then Y must be equivalent to X — Symmetric.
   iii) If X is equivalent to Y, and Y is equivalent to Z then X must be equivalent to Z — Transitive.
Does the definition of equivalent processes given in this section satisfy these three criteria?

3. IDENTITIES AND INVERSES

One permutation that is of particular interest is the one which leaves every piece alone in its home position. This is called the *identity permutation* and a process which produces the identity permutation is called an *identity process*. There are many processes which result in the identity permutation, as we demonstrate every time we restore the cube. A simple method of generating an identity process is to turn a face one way and then turn it back again, or to rotate it all the way around, 360°. A simple extension of this is to apply a sequence of turns and then reverse each turn in the opposite order. For example FR, followed by $R^{-1}F^{-1}$ is a trivial identity process. Notice that $FRF^{-1}R^{-1}$ is *not* an identity process.

For any process, we define its *inverse* to be the reverse of each turn in the process sequence applied in the opposite order. It is important that the reverse sequence be applied in the reverse order. An amusing illustration of the importance of the order for an inverse process is to consider two processes P and Q, where

$$P = \text{putting on your socks}$$

and

$$Q = \text{putting on your shoes}$$

Clearly the inverse of PQ must be $Q^{-1}P^{-1}$, not $P^{-1}Q^{-1}$, since taking off your socks before your shoes is decidedly unreasonable. Thus $Q^{-1}P^{-1}$ is the inverse of the process PQ. Any process followed by its inverse produces the identity permutation. The permutation produced by the inverse of a process is defined as the *Inverse permutation* of the process.

There are many other processes which are equivalent to the inverse of a process. That is to say, the inverse of a process is not the only way to produce the inverse permutation. For example, in Step 5 of the restoration method of the last chapter, one process described for relocating corners was

$$LD^2L^{-1}U^2LD^2L^{-1}U^2$$

which produces the permutation

$$(ufl, ubr, bdr).$$

The inverse permutation is

$$(ufl, bdr, ubr).$$

This can be produced either by

$$U^2LD^2L^{-1}U^2LD^2L^{-1}$$

which is the inverse of the original process or by the process

$$B^{-1}D^{-1}BU^2B^{-1}DBU^2.$$

Notice that the inverse of an inverse permutation is the original permutation. Therefore, two processes are equivalent if and only if their inverse processes are equivalent. In the example just given, this shows that $LD^2L^{-1}U^2LD^2L^{-1}U^2$ is equivalent to $U^2BD^{-1}B^{-1}U^2BDB^{-1}$.

In what follows it will sometimes be convenient to refer to an entire process by a single symbol such as

$$X=LD^2L^{-1}U^2LD^2L^{-1}U^2$$

and

$$Y=U^2B^{-1}D^{-1}BU^2B^{-1}DB.$$

We use $X=Y$ to denote that X is equivalent to Y. The notation X^{-1} refers to the inverse of the entire process X. The symbol I will be used for the identity process and the identity permutation. Thus we will write

$$XX^{-1}=I=X^{-1}X.$$

The concepts of identities and inverses occur frequently throughout mathematics and will reoccur several times through this book. In the next section we will examine how they are generated by the repetition of a single process.

EXERCISES:

4.3-1   For each process in Column A find the process in Column B which produces its inverse permutation.

	Column A		Column B
a)	$UDLRU^2D^2LRUDF^2B^2$	i)	$U^{-1}F^{-1}UBU^{-1}FUB^{-1}$
b)	$(F^2R^2)^2=F^2R^2F^2R^2$	ii)	$U^2R^{-1}L^{-1}URLU^2R^{-1}L^{-1}$
c)	$URU^{-1}L^{-1}UR^{-1}U^{-1}L$	iii)	$F^2B^2U^2F^2B^2R^2L^2(F^2U^2)^3R^2L^2(B^2U^2)^3$
d)	URL	iv)	$UDLRU^2D^2LRUDF^2B^2$
e)	D^2	v)	$F^{-1}B^{-1}L^{-1}R^{-1}U^{-1}D^{-1}$
f)	$UDLRFB$	vi)	$(F^2R^2)^4=F^2R^2F^2R^2F^2R^2F^2R^2$

4.3-2   Of the following statements, which are true and which are false?

a)   For each process there is one and only one process which produces the inverse permutation of the process.

b)   The permutation produced by a process followed by its inverse process is the identity permutation.

c)   The permutation of a process is always the same as the permutation of the inverse of that process.

d)   For each permutation there is one and only one inverse permutation.

e)   The inverse of UR is $U^{-1}R^{-1}$.

f)   The inverse of the identity is the identity.

g)   For processes X, Y, and Z, if YX is equivalent to ZX then Y is equivalent to Z.

4.3-3   a)   What is the inverse permutation of (A, B)? of (A, B, C)? of (A, B, C, D)?

b)   What is the inverse of a cycle?

c)   What is the inverse of (A, B) (C, D, E)?

d)   Can you state a rule for finding the cycle representation of a permutation given the cycle representation of its inverse?

4.   CYCLIC ORDER OF A PERMUTATION

The example of two equivalent processes given in Section 2 of this chapter demonstrates an interesting phenomonon. We observed that

$$F^2B^2L^2R^2$$

and

$$R^2L^2B^2F^2$$

are equivalent. But, notice that $R^2L^2B^2F^2$ is also the inverse of $F^2B^2L^2R^2$. Since $F^2B^2L^2R^2$ is equivalent to its own inverse, $(F^2B^2L^2R^2)(F^2B^2L^2R^2) = (F^2B^2L^2R^2)^2 = I$. But when we look at the permutation produced by $F^2B^2L^2R^2$, namely

(ufl, ubr)
(urf, ulb)
(dlf, drb)
(dfr, dbl)
(uf, df)
(ul, dl)
(ub, db)
(ur, dr),

we can see why that process is equivalent to its inverse. The permutation is made up of nothing but exchanges of pairs of pieces. So, repeating the process — thus repeating the permutation — just exchanges the pieces back again. No twists or flips occur, so exchanging the pieces back again produces the identity.

But, what would happen if, instead of pair exchanges, three pieces exchanged places as in

$$X = L^{-1}D^2LU^2L^{-1}D^2LU^2$$

which produces the permutation

(ulb, urf, frd)?

Then X^2 produces the permutation

(frd, urf, ulb)

which is the permutation of X^{-1}. Therefore we have

$$X^3 = XX^2 = XX^{-1} = \ .$$

In general, any single cycle of a permutation produces the identity when repeated as many times as there are places in that cycle. The number of places in a cycle is called the *length* of the cycle.

What happens when the permutation has cycles of different lengths? For example, the process $Y = F^2R^2$ produces the permutation

> *(ufl, ubr, dfr)*
> *(dlf, drb, urf)*
> *(uf, df)*
> *(ur, dr)*
> *(fl, br, fr).*

The cycles with two elements produce the identity on the cubicles in those cycles whenever Y is repeated twice and the cycles with three elements produce the identity on the cubicles in their cycles whenever Y is repeated three times. Therefore, repeating Y six times produces the identity for all cubicles, that is

$$Y^6 = I.$$

Incidentally, Y^3 is a useful process itself for moving edges without moving corners as we will discuss in the next section.

When cycles include flips or twists you must be careful to include the flips or twists in the length of the cycle. For example,

$$Z = FR^{-1}$$

produces the permutations

> *(ufl, rdf, dlf)_*
> *(urf, rbd, rub)*₊
> *(uf, rd, rb, ru, rf, df, lf).*

The twists on the corner cycles multiply the length of those cycles by three. Since Z^3 does not result in an identity on the corner pieces but only returns them to their home locations twisted, the lengths of the cycles are *not* 3, 3, and 7, but rather are 9, 9, and 7. If Z is repeated 21 times, all pieces will be in their home locations, but the six corners will be twisted. We find that the smallest number of repetitions of Z which gives the identity is 63.

By now you should see that every process if repeated enough times will result in the identity. The smallest number of times that a process must be repeated to produce an identity on the entire cube is called the *order of the process*. For example, we have shown that the order of the process $F^2B^2L^2R^2$ is two. The processes

$$X = L^{-1}D^2LU^2L^{-1}D^2LU^2$$
$$Y = F^2R^2$$
$$Z = FR^{-1}$$

we have shown to have order 3, order 6, and order 63 respectively. Since the order of a process is determined by the permutation produced by that process, then equivalent processes have the same order. That same number is also called the *order of the permutation* of that process. Thus the order of a permutation is the least common multiple — denoted LCM — of the orders of its cycles.

EXERCISES: (* indicates harder.)

4.4-1   What is the order of each of the following?
   a)   U
   b)   FR
   c) FR^{-1}
   d) $FRF^{-1}R^{-1}$
   e) LD^2L^{-1}
   f) $U^2LB^{-1}D^2BL^{-1}$

4.4-2   Find a process of order three which moves only edge pieces.

4.4-3**   What is the largest order that any process on the cube can have? Give an example of such a process.

5. FINDING USEFUL PROCESSES

It is interesting to see what happens when a process is repeated enough times to produce the identity for some cycles of its permutation but not for all the cycles. Again, for example, consider

$$Z = FR^{-1}$$

whose permutation is

*(ufl, rdf, dlf)_*
(urf, rbd, rub)_+
(uf, rd, rb, ru, rf, df, lf).

If we repeat Z seven times then

$$Z^7 = FR^{-1}FR^{-1}FR^{-1}FR^{-1}FR^{-1}FR^{-1}FR^{-1}$$

has a permutation

*(ufl, dfr, fdl)_*
(urf, drb, ubr)_+

which is similar to the process Z on the corners — not equivalent since the corner orientations are different — but leaves the edges fixed. Thus we have built a process for moving corners while leaving edges fixed. Unfortunately that moves too many corners to be useful very frequently.

A more useful example is provided by the process

$$Y = F^2R^2$$

which produces the permutation

(ufl, ubr, dfr)
(dlf, drb, urf)
(fl, br, fr)
(uf, df)
(ur, dr).

Then the permutation produced by Y^2 is

(ufl, dfr, ubr)
(dlf, urf, drb)
(fl, fr, br)

and the permutation produced by Y^3 is

(uf, df)
(ur, dr).

The latter of these is particularly useful in moving edge pieces without disturbing corners. Although this is not required in the restoration method described in Chapter 3, it is a useful process for solving many problems quickly. Also, in Step 3 of the restoration method, pieces which are in the middle slice but in the wrong location can sometimes be moved directly to their home position by $\mathcal{J}Y^3\mathcal{B}$ without moving the piece to the Up face. Further, there are some times when many corners happen to be in place when it is useful in the restoration method.

Another example of a useful process is

$$P = URU^{-1}R^{-1}$$

whose permutation is

$$(ulb, ubr)_+$$
$$(urf, frd)_-$$
$$(ub, ur, fr).$$

We will see more and more of this process as we proceed. For now it will be sufficient to notice that

$$P^3 = (URU^{-1}R^{-1})^3$$

produces the permutation

$$(ulb, bru)$$
$$(urf, dfr)$$

which can often be useful in Step 5 of the restoration procedure.

EXERCISES:

4.5-1   Find a process which produces a permutation consisting of a single cycle of seven edge pieces and leaves all corners in their home position.

4.5-2   What is the smallest power of each of the following processes which will produce a permutation which leaves all edges in their home positions?

a) FR^{-1}
b) $F^2U^{-1}DR^2UD^{-1}$
c) LF^2L^{-1}
d) $UFRF^{-1}R^{-1}U^{-1}$
e) $FR^{-1}F^{-1}RF^{-1}U^{-1}F$
f) $R^{-1}FRF^{-1}U$
g) $R^2U^2F^2$

6. COMMUTATIVITY AND COMMUTATORS

The process $P = URU^{-1}R^{-1}$ discussed in the previous section shows several interesting points. First, we have already observed that $U^{-1}R^{-1}$ is *not* equivalent to the inverse of UR. The inverse of UR is $R^{-1}U^{-1}$. The sequence of applying the processes makes a difference. To clarify this difference for yourself, you should use your cube to derive and compare the permutations produced by the two processes UR and RU.

For the numerical operations of addition and multiplication, the order of the numbers does *not* make a difference. Thus $3+5$ equals $5+3$ and $3\cdot5$ equals $5\cdot3$. Operations where the order of the operands does not matter are called *commutative*. Operations where the order does matter are called *non-commutative*. Subtraction and division are non-commutative. For example $3-5$ does not equal $5-3$. The combining of processes on the cube is non-commutative. For instance, the above example shows that UR does not equal RU. The order in which the processes are applied makes a big difference.

Although $URU^{-1}R^{-1}$ is not the identity, it is a lot less complex than UR alone. The permutation of UR is

$(ubr)_+$
$(urf, ufl, ulb, bdr, dfr)_-$
$(uf, ul, ub, br, dr, fr, ur).$

The permutation of $U^{-1}R^{-1}$ is

*(urf)_*
(ubr, ulb, ufl, frd, drb)_+
(uf, fr, dr, br, ur, ub, ul).

Many of the pieces that are moved by UR are returned to where they started by $U^{-1}R^{-1}$. If a piece is moved by U to a place that is not moved by R, then it will be moved back by U^{-1} to where it started. And, if that place where it started is not moved by R^{-1} — or equivalently, is not moved by R — then the process $URU^{-1}R^{-1}$ ends up leaving the piece where it started. Similarly, pieces that are moved by R to and from places which are not moved by U are left where they started by $URU^{-1}R^{-1}$. Only where there is overlap between the processes are pieces affected. Thus the permutation produced by $URU^{-1}R^{-1}$ only affects pieces which move to or from locations common to both the Up and Right faces. Thus the permutation is

(ulb, ubr)_+
*(urf, frd)_*
(ub, ur, fr)

where *ubr, urf,* and *ur* are the three locations common to both faces.

Many useful processes are formed in a similar way, exploiting the fact that $XYX^{-1}Y^{-1}$ is not the identity. Any process having the form $XYX^{-1}Y^{-1}$ is called a *commutator*. Commutators appear useful so often in mathematics wherever non-commutative operations occur that a shorthand notation has been accepted for writing a commutator of any two elements. Namely the commutator of two elements, X and Y is written

$$[X, Y] = XYX^{-1}Y^{-1}.$$

Thus the commutator $URU^{-1}R^{-1}$ is written [U,R] while the commutator $UR^{-1}U^{-1}R$ is written [U,R^{-1}]. Uses for these two commutators on the cube in particular occur so frequently and are so powerful that they have been given nicknames, the *Z commutator* and the *Y commutator* respectively. These names come from the pattern of the locations moved by each as shown in Figures 4-3 and 4-4. The names, Z commutator and Y commutator are used regardless of which two adjacent faces are used in the commutator. Thus $[R,F] = RFR^{-1}F^{-1}$ is a Z commutator and $[R,F^{-1}] = RF^{-1}R^{-1}F$ is a Y commutator even though on the latter the cube must be turned over to see the Y pattern.

THE Z COMMUTATOR

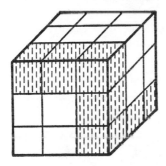

Shading indicates the locations changed by $FRF^{-1}R^{-1}$.

Figure 4-3

THE Y COMMUTATOR

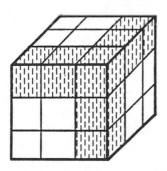

Shading indicates the locations changed by $FR^{-1}F^{-1}R$.

Figure 4-4

The Z and Y commutators are so powerful that you could use either of them, applying each to different pairs of faces and restore any scrambled cube without using any other processes. That would be a long and cumbersome method but it can be done. In the next section we will see how they can be particularly useful in moving corners without affecting edge locations.

The principle of commutators was used in Chapter 3 with U and R replaced by more complex processes. Examples of such processes of the form $[X,Y] = XYX^{-1}Y^{-1}$ occur in Step 5 and Step 6 of the restoration method. To move *ufl* to *urf* let

$$X = LD^2L^{-1}$$

and

$$Y = U.$$

Then the commutator

$$XYX^{-1}Y^{-1} = LD^2L^{-1}ULD^2L^{-1}U^{-1}$$

produces the permutation

(ufl, urf, bdr).

The only location of the cube which is affected by both X and Y is the corner location *ufl*. Therefore, only the pieces moved to or from *ufl* by either X or Y are moved by the commutator $XYX^{-1}Y^{-1}$.

To untwist corner pieces in Step 6, let

$$X = L^{-1}D^2LBD^2B^{-1}$$

and

$$Y = U.$$

Then the commutator

$$XYX^{-1}Y^{-1} = L^{-1}D^2LBD^2B^{-1}UBD^2B^{-1}L^{-1}D^2LU^{-1}$$

produces the permutation

(ulb)_ (ufl)_+

which is a counter-clockwise twist of *ulb* and a clockwise twist of *ufl.* In this example the only location of the cube which is affected by both X and Y is the corner location *ulb*, and X does not move it to another location, but only twists it in place. Then by applying U followed by X^{-1}, we not only restore the Down and middle layers but untwist, in place, the piece moved from *ufl* into *ulb* by U. Finally U^{-1} moves the piece that started in *ufl* back to *luf* and the piece that started in *ulb* back to *lbu.*

Sometimes the orientation and notation which we have chosen, can obscure a process from being seen as a commutator. For example, the process

$$W = U^2 L R^{-1} F^2 L^{-1} R$$

is not seen as a commutator in our orientation and notation. But if we do not require that the center pieces of each face stay fixed, then we can use the moves which reorient the entire cube. This enables us to see the process as a commutator. We can write

$$W = U^2 L R^{-1} \mathcal{R} U^2 \mathcal{R}^{-1} R L^{-1}$$
$$= [U^2, L R^{-1} \mathcal{R}]$$

which is clearly a commutator. In either form the process W produces the permutation

(ub, df, uf)

which is useful for moving edges without disturbing corners.

Sometimes a commutator $XYX^{-1}Y^{-1}$ is equal to the identity I. However we have

$$XYX^{-1}Y^{-1} = I$$

if and only if

$$XY = YX.$$

In this case we say X and Y *commute*. One condition that is sufficient to assure that $XY = YX$ is that the cycles of X and of Y are disjoint. There are other cases where X and Y also commute. For instance they may have one or more cycles which are identical and their other cycles may be disjoint. A cycle of X may move pieces through the same locations as a cycle of Y and give every piece in the cycle a similar twist or flip. Thus we can come up with many examples of processes X and Y which do commute — equivalently for which $XYX^{-1}Y^{-1} = I$ — but in the vast majority of cases the processes do not commute.

EXERCISES: (* indicates harder.)

4.6-1** Find a process Q other than the identity which commutes with every other process on the cube, that is

$$QX = XQ$$

for every process X on the cube.

4.6-2 Let S_X denote the set of locations which are affected by the process X and let S_Y denote the set of locations which are affected by the process Y.

a) If S_X and S_Y have no locations in common, what is the permutation of $XYX^{-1}Y^{-1}$?

b) If S_X and S_Y have two locations in common, what is the largest number of locations which can be in S_Z where $Z = XYX^{-1}Y^{-1}$?

c) If S_X and S_Y have two locations in common, what are all the numbers that could be the number of locations in S_Z?

4.6-3 What is the permutation produced by $[F,R^{-1}][R,U^{-1}][U,F^{-1}]$?

4.6-4 Show that
   a) $[F,R]^{-1} = [R,F]$
   b) $[F,R^{-1}]^5 = R^{-1}[F,R]R$
   c) $[F^{-1},R^{-1}] = R^{-1}F^{-1}[F,R]FR$

4.6-5 What is the permutation produced by $[(R^2U^2F^2)^3, U^2]$?

4.6-6 What is the inverse of $[X,Y]$?

7. CONJUGATES: BUILDING NEW PROCESSES FROM OLD

Whereas a commutator is useful in deriving new processes, a conjugate is useful in modifying known processes to apply to different pieces. A conjugate is a process of the form

$$XYX^{-1}.$$

This is called the conjugate of Y by X. Many of the processes used in the restoration method were shown in Chapter 3 to be examples of conjugates.

Consider the process

$$RUR^{-1}$$

which produces the permutation

$$(uf, ul, ub, fr)$$
$$(ufl, ulb, fur, frd).$$

Notice that the cycle structure of this permutation is the same as the cycle structure of the process U alone, namely a four-cycle of edges without a flip and a four cycle or corners without a twist. This is a characteristic of conjugates.

For another example let Y be the commutator

$$Y = L^{-1}D^2LUL^{-1}D^2LU^{-1} = [L^{-1}D^2L, U]$$

whose permutation is

$$(ulb, ufl, frd)$$

and let

$$X = R^{-1}.$$

Then the conjugate

$$XYX^{-1} = R^{-1}L^{-1}D^2LUL^{-1}D^2LU^{-1}R$$

produces the permutation

(ulb, ufl, urf).

Again the cycle structure of the permutation of the conjugate XYX^{-1} is the same as the cycle structure of the permutation of the process Y. These conjugates of the basic corner 3-cycles are particularly useful in Step 5 to minimize the twists needed in Step 6.

In general to use a conjugate of the form XYX^{-1}, the process X places pieces in the positions to be moved by the permutation of Y. Then Y moves the pieces. Then X^{-1} returns the now permuted pieces to their new places. The effect is to produce a permutation with the same cycle structure as Y but acting on different locations.

Let us examine this more carefully. Suppose Y carries the object at location a to b — that is, $Y(a) = b$ — and suppose X carried the object at location c to a and the object at location d to b. Consider the action of XYX^{-1} on $c = X^{-1}(a)$ as shown in the following diagram:

$$X^{-1}(a) = c \xrightarrow{X} a \xrightarrow{Y} b \xrightarrow{X^{-1}} d = X^{-1}(b).$$

So we have

$$XYX^{-1}(c) = d$$

or expressed another way we have

$$XYX^{-1}(X^{-1}(a)) = X^{-1}(Y(a)).$$

Thus we see that the permutation produced by XYX^{-1} has the same behavior on the location $X^{-1}(a)$ as the behavior produced by Y on the location a for all locations a on the cube. In particular, the cycle representation has the same form — this is called the *cycle structure*. The cycle structure consists of the number of cycles and the length of those cycles. For example, U consists of two 4-cycles. So any conjugate of U also consists of two 4-cycles.

You probably have already developed new applications of this principle of conjugates for yourself in untwisting corners in Step 6 of the restoration method. In particular, it is useful when the two corners to be untwisted are not on the same face. For example, to twist *ulb* clockwise to *lbu* and twist *dfr* counter-clockwise to *rdf*, we choose

$$X = R$$

and choose Y to be the commutator

$$Y = BD^2B^{-1}L^{-1}D^2LU^2L^{-1}D^2LBD^2B^{-1}U^2 = [BD^2B^{-1}L^{-1}D^2L,U^2].$$

Then XYX^{-1} produces the permutation

$$(ulb)_+ (dfr)_-.$$

Another conjugate will exchange the places of any three edge pieces without disturbing any other pieces of the cube. To do this we choose

$$W = U^2LR^{-1}F^2L^{-1}R$$

which we saw in the preceeding section produces the permutation

$$(ub, df, uf)$$

If, for example, we need a process to produce the permutation

$$(ub, ur, uf),$$

then all we need is to find a process, X, which leaves *ub* and *uf* in place and moves *ur* to *df*. What X does to the rest of the cube doesn't matter. So we choose

$$X = R^2D^{-1}$$

which has the permutation

$$(ur, df, dl, db, dr)$$
$$(fr, br)$$
$$(urf, dfr, ubr, dlf, dbl, drb).$$

Then the conjugate

$$XWX^{-1}$$

produces the permutation

(ub, ur, uf).

Another example which can be useful in Step 5 of the restoration method is obtained by choosing

$$Y=(URU^{-1}R^{-1})^3$$

which was seen earlier to produce the permutation

(ulb, bru)
(urf, dfr).

Then the conjugate

$$FYF^{-1}$$

produces the permutation

(ulb, bru)
(ufl, fur),

using fewer moves than the methods of Step 5 and producing different twists to the corners.

EXERCISES:

4.7-1   Find conjugate processes A_i and B_j for each of the permutations below. Then find a list of processes C_i such that each process B_j from Column B is shown to be the conjugate of A_i by the process C_i — that is, choose C_i such that $B_j=C_iA_iC_i^{-1}$

	Column A		Column B
1)	*(ufl, urf, bdr)*	1)	*(dfr, dlf)_ (urf, ufl)+*
2)	*(ulb, ubr)+ (ufl, urf)_*		*(ul, fd, uf)*
	(ub, ur, fu)	2)	*(ubr, urf)(lf, bl)*
3)	*(ufl, urf)(uf, ur)*	3)	*(ufl, dlf, fur)*

4.7-2   Show that $(XYX^{-1})^3=XY^3X^{-1}$.

4.7-3   The process $Y = (F^2R^2)^3$ produces the permutation
$$(uf, df) (ur, dr).$$
Find X such that the conjugate of Y by X produces the following permutations:

   a)   *(uf, df)(rf, rb)*
   b)   *(fu, fd)(fr, fl)*
   c)   *(fu, fr)(fb, fl)*
   d)   *(uf, rb)(ub, rb)*

Find X such that the conjugate of $Y = F^2RL^{-1}U^2R^{-1}L$ — with permutation *(fu, fd, bu)* — by X produces the following permutations:

   e)   *(fu, ru, bu)*
   f)   *(fu, ur, rf)*
   g)   *(fu, rb, dl)*

Find X such that the conjugate of $Y = [U,R]^3$ — with permutation *(ulb, bru) (urf, dfr)* — by X produces the following permutations:

   h)   *(ulb, bru) (ufl, fur)*
   i)   *(ulb, urf) (ufl, bru)*
   j)   *(ulb, urf) (dlf, drb)*

4.7-4   Find a process which produces the permutation
$$(lf, bl, rb, fr)$$
$$(dlf, dbl, drb, dfr).$$

CHAPTER 5

IMPROVED RESTORATION PROCESSES

We have already seen some examples of how commutators and conjugates are used in the restoration method of Chapter 3. However, a more thorough look will give many more examples showing how we can use permutation cycles, repetitions of permutations, commutators, and conjugates to improve the method further.

1. THE DOWN-FACE EDGES

We can see that each Down-face edge piece can be restored in at most four moves — see Section 1 of Chapter 3 — and therefore at most 16 moves are needed to complete this step. The simplicity of restoring the first piece will reduce this by a few moves, but if it is to be substantially reduced, one must consider more than one piece at a time. This requires using permutation cycles to place several pieces in their home positions simultaneously.

Since the effects of the permutations on corners is irrelevant at this step, there are more potentially useful processes than those which can be listed here. It has been shown that this step can be done in 12 moves but it has not yet been determined what is the minimum number of moves which will be sufficient to complete Step 1 from any starting position.

A few simple conjugates and commutators are listed here which may be useful in some cases and give you more ideas. You should develop for yourself such a larger list as you find useful.

$F^2B^2U^2B^2F^2$ → *(df, db)(ul, ur)(dlf, drb)(dfr, dbl)*

$DRD^{-1}R^{-1}$ → *(df, dr, br)(dlf, dfr)$_+$(drb, bru)_*

$F^{-1}RFR^{-1}$ → *(df, rf, rd)(dlf, bdr)$_+$(dfr, rfu)_*

$RBF^{-1}R^{-1}B^{-1}F$ → *(df, rd, rf)(db, ru, rb)(dlf, dbl, drb)_*
(dfr, fur, ubr)$_+$

2. THREE DOWN-FACE CORNERS

We noticed in Chapter 3 that to move a corner piece into its home position on the Down face, it was easier if the piece was on the Up layer with its Down-face color on one of the sides, not on the Up face. In this case there are six different three-move conjugates which might be applied, only two of which were used in Chapter 3. For example, to move the *DLF* cubie from a Up-face corner to its home location of *dlf*, you can use any one of the following:

FU^2F^{-1} → *(rub, dlf)(ulb, luf)(lf, ub)(ul, ur)*

$L^{-1}U^2L$ → *(bru, dlf)(urf, flu)(lf, ru)(uf, ub)*

FUF^{-1} → *(flu, dlf, bul, rub)(lf, ul, ub, ur)*

$L^{-1}U^{-1}L$ → *(luf, dlf, rfu, bru)(lf, fu, ru, bu)*

$FU^{-1}F^{-1}$ → *(bul, dlf, flu, rub)(lf, ur, ub, ul)*

$L^{-1}UL$ → *(rfu, dlf, luf, bru)(lf, bu, ru, fu).*

Rotating the Up face in order to use a different one of these can sometimes make restoring the next corner more simple. For example,

$$U^2FUF^{-1} \rightarrow (bru, dlf, bul, fur)_+ (ufl)_- (lf, ul)(uf, ur, ub)$$

moves *bru* to *dlf* just as does $L^{-1}U^2L$. But the two processes have very different effects on the other Up-face corners which may be used to simplify the next corner move. If the piece which started in the *dlf* location also needs to be moved to the Up face, preparatory to being moved to a Down-face home location, then you want to avoid ending up with the Down-colored facelet on the top. Choosing U^2FUF^{-1} moves the Left facelet of the *dlf* location to the Up face whereas choosing $L^- U^2L$ moves the Front facelet of the *dlf* location to the Up face. Thus the sequence of restoring the corner pieces and the choice of processes can be used to minimize the number of moves required for Step 2 even when each of the three corners is restored separately.

A more sophisticated approach is to restore several corners simultaneously. For example to move

$$ulb \rightarrow dlf$$

and

$$ufl \rightarrow dbl$$

use $FB^{-1}U^2F^{-1}B$. This incidentally is a simple way of restoring either of these corners when the desired Down-colored facelet is on the Up face and the other corner has not already been restored. Variations similar to this for moving two pieces at once from the Up layer are given by processes which first move several Down-face edges to the middle layer — none to the Up face — second, rotate the Up face, and finally restore the Down-face edges. These processes are all conjugates of the Up-face rotation. Thus the permutation of corners for any of these processes is either a 4-cycle — for conjugates of U or U^{-1} — or is two 2-cycles — for conjugates of U^2. Therefore, no more than two Up-face corner cubies can be moved to two Down-face corner cubicles in a single process of this type — since only four corners are involved in the permutation of the process. However, more than two corners can be restored simultaneously using such a conjugate by also moving a Down-face corner to another Down-face corner. For example, the process $RL^{-1}D^2BUB^{-1}D^2R^{-1}L$ produces the permutation

(rdf, dlf, fur, dbl)(lf, ur, rf, fu).

Can you find a process which produces a 4-cycle of the Down-face corners without moving the Down-face edges?

3. MIDDLE-LAYER EDGES

Middle-layer edges can also benefit from processes that can restore several edges at once. Of particular value are processes which move edges within the middle layer. An excellent example was given — in a different orientation — in the earlier discussion of commutators, namely,

$$F^2UD^{-1}L^2U^{-1}D$$

which produces the permutation

(lf, rf, lb).

The process $(U^2R^2)^3$ which produces the permutation

(uf, ub)
(rf, rb)

may also be useful, as well as the other orientations of this process and their conjugates, such as $F(U^2R^2)^3F^{-1}$. Some other potentially useful processes are given in Appendix A, but the simplicity of the middle layer restoration makes it difficult to substantially reduce the number of moves required in general.

Some forethought during Step 3 may be useful in reducing the requirements for Step 4. In particular when edge pieces are removed from the middle layer, it may be possi-

ble to place them in their ultimate home positions in the Up
face, or at least with their Up-colored facelet on the Up face.

4. FINAL UP-FACE EDGES

To reduce the number of moves required in this step you
again should look to processes which will restore several
cubies simultaneously to their home positions. A potentially
useful collection of processes is obtained from conjugates
of the Z and Y commutators of two faces discussed in Section 6 of Chapter 4. These commutators all permute three
edge pieces cyclically and exchange two pairs of corners.
Notice that the basic moves used in Step 4 can be seen to
include these processes by letting U and U^{-1} precede the
conjugates of U^{-1} and U respectively. However, conjugates
of these commutators are not included. For example, consider the process

$$FRUR^{-1}U^{-1}F^{-1}$$

which is the conjugate of $RUR^{-1}U^{-1}$ by the rotation F and
produces the permutation

$$(uf, ru, bu)$$
$$(urf, ufl)_+$$
$$(ubr, ulb)_-.$$

To see how many moves that process or others of that type
might save you, take a restored cube and apply the inverse
of the process above, namely,

$$FURU^{-1}R^{-1}F^{-1}.$$

Now, count the number of moves it takes you to put the
edges back in place using the techniques you have been
using for Step 4. Sure, it messes up the corners. But, if you
had been restoring a scrambled cube, for all you know,
either process might have fixed them.

If the completion of Step 4 requires a cyclic rotation of
three edges, these conjugates of Z and Y commutators can

frequently be used. However, a cyclic rotation of four edges without affecting other edges is difficult. As we saw in Section 4 of Chapter 3, that permutation always necessitates at some point making a quarter turn of one *center piece* relative to the home locations of the edges on that face. We will discuss why this is so in Chapter 7. But, for now it is sufficient to observe that when choosing the home location for the first Up-face edge to be restored in Step 4, you should avoid creating the need for a cyclic rotation of all four of the remaining unrestored edges. A cyclic rotation of three edges or the exchange of two pairs of edges will be much simpler — that is, require fewer moves.

5. RESTORING CORNERS UNTWISTED

It is in Step 5 that the greatest savings can be made. In particular, by choosing the processes carefully in Step 5, you can usually eliminate the need for any of the untwisting of Step 6 — that is, eliminate Step 6 altogether. First, notice that the processes described in Step 5 for moving corners, namely

$$F^{-1}D^2FU^iF^{-1}D^2FU^{-i}$$

and

$$LD^2L^{-1}U^iLD^2L^{-1}U^{-i}$$

where $i = 1, 2,$ or -1, move the selected corner cubie to the target corner with the same facelet showing on the Up face. Therefore choosing the selected corner to be a piece which already has the Up-colored facelet showing on the Up face will eliminate the need to reorient that piece after placing it in its target corner. If none of the Up-face corners to be moved have the Up-face color showing on the Up face, then perhaps one of them has a side facelet color matching the color of that side center piece. In that case, reorienting the cube to put that side on top will allow you to use the same processes for moving the selected corner to the target corner.

Either of the two processes above will move the selected corner to the target corner maintaining the Up-facelet color. But, by your choice between the two processes, you will determine the orientation of the working corner when it moves to the selected corner, and the orientation of the target corner when it moves to the working corner. In particular, the process

$$F^{-1}D^2FUF^{-1}D^2FU^{-1}$$

produces the permutation

(ufl, urf, rbd)

and the process

$$LD^2L^{-1}ULD^2L^{-1}U^{-1}$$

produces the permutation

(ufl, urf, bdr).

Thus if the home location of the target corner is the working corner or the home location of the working corner is the selected corner, then carefully choosing which process to use can often restore one of those with the proper orientation as well.

Two applications of these processes can usually be used to restore with the proper orientation at least two corners of the five involved in Step 5 while leaving the other three requiring a 3-cycle corner permutation. It is easier to restore those last three corners if they need a 3-cycle permutation than it is to twist them within their home locations. Therefore, the first two processes should be chosen, not only to restore two corners, but also to move corners which are already in their home locations but not in their home positions. Try to avoid leaving corners twisted in their home locations. You can do this while restoring two corners unless Step 5 started with all five corners in their home location but all incorrectly oriented.

Can you figure out how many different ways the five corners being restored in Step 5 can all be in their home locations but twisted with the other three Down-face corners all
being in their home positions? This is not easy but you may
enjoy guessing. To prove your answer you would need the
material of Chapter 7.

6. CONJUGATES HELP ORIENT THE FINAL CORNERS

To generate a 3-cycle permutation which will move the
last three corners all to their home positions and not just to
their home locations, we can often use a simple conjugate
of the commutator process we've been using. For example,
to produce the permutation

(ufl, fur, rub)

use the process

$$RF^{-1}D^2FU^2F^{-1}D^2FU^2R^{-1}.$$

Or, for a less obvious example, to produce the permutation

(ufl, urf, ulb)

use the process

$$L^2F^2LBL^{-1}F^2LB^{-1}L$$

which is seen to be such a conjugate when written

$$L^{-1}(L^{-1}F^2LBL^{-1}F^2LB^{-1})L.$$

The process inside the parenthesis is one of the same processes we have been using with the entire cube reoriented
by $\mathcal{L}\mathcal{U}$ to place the *bld* corner in the *ufl* position.

Earlier in this chapter while discussing Step 2 we listed
six processes for moving a corner from the Down layer to
the Up layer. So far we have been using the commutators of
only two of these six processes, namely $[U^i, LD^2L^{-1}]$ and
$[U^i, F^{-1}D^2F]$ where $i = 1, 2, -1$. The other four can also be

used to advantage in some cases. For instance to obtain the permutation

(ulb, ufl, bld)

use the process

$$UF^{-1}DFU^{-1}F^{-1}D^{-1}F$$

or to obtain the permutation

(ulb, rub, ufl)

reorient the entire cube using $\mathcal{BL}$ so that the *ulb* corner is in the *lbu* position and use the same process.

The Z and Y commutators can also be very useful for restoring those corners. Many people familiar with these commutators use them exclusively for Steps 5 and 6. Let us consider several examples.

In the last chapter we saw one example of such a process. The cube of the Z commutator

$$(URU^{-1}R^{-1})^3 = [U,R]^3$$

produces the permutation

(ulb, bru)
(urf, dfr).

Simple conjugates of this process can be very useful in exchanging pairs of corners. For example, the process

$$F(URU^{-1}R^{-1})^3F^{-1} = F[U,R]^3F^{-1}$$

produces the permutation

(ulb, bru)
(urf, flu)

in 14 moves. Another more sophisticated application of the Z commutator is the process

$$(UBU^{-1}B^{-1})F^{-1}(UBU^{-1}B^{-1})F(UBU^{-1}B^{-1})$$

which produces the permutation

$$(urf, ufl)\_$$
$$(ubr, drb)$$
$$(ulb)_+.$$

We will see more examples using these Z and Y commutators in the next section.

Another process for interchanging two pairs of corners is a variation of the process used to produce 3-cycles of the corners. Consider the process

$$FL^{-1}D^2LF^{-1}UFL^{-1}D^2LF^{-1}U^{-1}$$

which produces the permutation

$$(ulb, fur)$$
$$(ufl, rub).$$

This is the commutator

$$[FL^{-1}D^2LF^{-1},U]$$

the first part of which is the conjugate of D^2 by FL^{-1}. Many useful corner pair exchanges can be obtained from processes of this type.

7. UNTWISTING CORNERS

Although Step 6 can often be eliminated by careful performance of Step 5, it is still useful to be able to do twists efficiently. We observed in the last chapter how a conjugate can be used to extend the process for twisting and untwisting two corners on a single face, so that we can untwist two corners which have no face in common. Namely, if a process, X, has the permutation

$$(ulb)_+$$
$$(urf)\_$$

then the process

$$RXR^{-1}$$

has the permutation

$$(ulb)_+$$
$$(frd)_-.$$

To produce three corner twists in the same direction, the Z or Y commutators are particularly useful because their only edge permutation is a 3-cycle. To make use of this, we observe that

$$([R,D]^2)^3 = I.$$

The next important observation is that $[R,D]^2$ affects only the *ubr* location on the Up face, giving it a clockwise twist. If we rotate the Up face between three applications of $[R,D]^2$, we can twist three Up-face corners clockwise while producing an identity on the rest of the cube. For example, the process

$$([R,D]^2U)^3U = (RDR^{-1}D^{-1})^2U(RDR^{-1}D^{-1})^2U(RDR^{-1}D^{-1})^2U^2$$

produces the permutation

$$(ubr)_+ \ (urf)_+ \ (ufl)_+.$$

The 3-cycle permutations also can be used to twist three corners in the same direction. For example, the process

$$BD^2B^{-1}UBD^2B^{-1}U^{-1}R^{-1}B^2RF^2R^{-1}B^2RF^2 =$$
$$[BD^2B^{-1}, U] \ [R^{-1}B^2R, F^2]$$

produces the permutation

$$(ulb)_- \ (ufl)_- \ (frd)_-.$$

Before ending this chapter, we want to take care to point out that *not all* of the most efficient processes have been constructed by building with simple conjugates and commutators. For example, an English mathematician named Morwen Thistlethwaite uses the process

$$LBL^{-1}B^{-1}U^2F^{-1}L^2FL^{-1}F^{-1}L^2FL^{-1}U^2$$

to produce the permutation

$$(ulb)_- \ (ubr)_- \ (urf)_-$$

and Morwen Thistlethwaite's computer has generated the process

$$B^{-1}U^2B^2UB^{-1}U^{-1}B^{-1}U^2FRBR^{-1}F^{-1}$$

to produce the permutation

$$(ub)_+ \ (ur)_+.$$

These are the shortest known processes for these permutations. Still more processes are listed in Appendix A.

CHAPTER 6

THE CUBE GROUP
AND SUBGROUPS

We have seen that each process on the cube generates a permutation of the pieces of the cube. Further, we have seen that if one process on the cube is followed by another process, then the two combined form a new process which generates another permutation of the cube. Although this may seem trivial to you, to experienced mathematicians it is significant. Being able to combine two objects to form another object of the same set is the first requirement for a *group*.

The concept of a group appears in different branches of mathematics and has many applications in science and art. In particular, group theory is the mathematical foundation of the study of symmetry which is important in geometry, art, physics, chemistry, and biology. Group theory includes the study of permutations which are basic in coding theory, cryptography, English bell-ringing, magicians' card-shuffling tricks, etc. The rigid motions of space form a group and the study of such groups is basic in theoretical physics.

Group theory also turns out to be a way to study the solvability of polynomial equations and the structure of geometric and topological objects. Consequently, group theory has become one of the basic subjects of mathematics. We shall see that the processes and permutations on the cube form groups. These groups are useful concrete examples of this important theory, especially since the structure of these groups provides a concrete embodiment of many concepts in group theory that students find difficult to grasp without physical examples. The cube can literally be grasped!

1. THE PERMUTATIONS OF THE CUBE FORM A GROUP

What is a group? To start with, a group consists of two things, a set of objects — denoted by S — and an operation — denoted by * — which combines two of these objects together to form another object in S. Formally we should always refer to a group (S,*), but usually the operation is clear in context. In those cases, we will refer simply to the group S.

For examples of a group, the set of objects S could be numbers and the operation for combining them could be addition or multiplication. In the case of the cube, the set of objects is the permutations on the cube and the operation for combining them is "followed by". Thus, if X and Y are permutations on the cube, then X and Y are combined to form a single permutation

X followed by Y

which is written simply

XY.

In order for a set of objects S and an operation * to form a *group*, they must meet four criteria.

1. *Closure Law:* Whenever two objects X and Y are in the set S then the result of combining X and Y by the operation

* must be another object in the set S — that is X*Y is in S. When X and Y are permutations on the cube then "X followed by Y" is another permutation on the cube. Thus this criterion is met.

  2. *Associative Law:* Whenever three objects X, Y, and Z are combined in a fixed order then (X*Y)*Z produces the same result as X*(Y*Z).

Technically, the parentheses are needed since the operation only combines two objects, but the Associative Law says that in a group, parentheses really don't matter. If they do matter, then the operation cannot be used to form a group. Notice that combining processes or permutations on the cube by using the operation "followed by" meets this criterion — but you must be careful not to change the order of the processes.

  3. *Identity Law:* There is a unique object in the set S which is called the identity of the group and is denoted by I. It has the property that for every X in S

$$I*X = X*I = X.$$

So, combining the identity with any object X in the set produces the object X again. The permutation which does not move any pieces is, as we have seen, the identity for the set of permutations on the cube. The process of not turning any face, that is of doing nothing is the identity for the set of processes on the cube.

  4. *Inverse Law:* For each object X in the set S there is a unique object in S called the inverse of X, which is denoted by X^{-1} and has the property that

$$X*X^{-1} = X^{-1}*X = I.$$

That is, either combining X with X^{-1} or X^{-1} with X produces the identity object. We saw in Chapter 4 that each process on the cube has an inverse process and its permutation is the inverse of the permutation of the original process.

A set of objects and an operation for combining pairs of objects in the set form a group if the above four criteria are met. Thus the processes and the permutations on the cube form groups. We actually have two groups. In the group of processes, we consider F, FF = F^2, FFF = F^3, F^4, F^5, ... as distinct processes — which they are — and so, using turns of all six faces, we have an infinite group which group theorists call the *free group* on six generators. However, we are really interested in the permutations produced by the processes. Then we have $F^4 = I$, $F^5 = F$, etc. and we have a finite group of permutations which we call the *Cube Group*.

EXERCISES:

6.1-1   Which of the four criteria for a group are satisfied by the set S of positive real numbers, combined by each of the following operations:
  a. addition — that is, X∗Y is X+Y?
  b. multiplication — that is, X∗Y is X·Y?
  c. division — that is, X∗Y is X ÷ Y?
  d. subtraction — that is, X∗Y is X − Y?
  e. maximum — that is, X∗Y is the larger of X and Y?
  f. average — that is, X·Y is (X+Y)/2?
  g. last — that is, X∗Y is Y?

6.1-2   Which of the four criteria for a group are satisfied by the set of four permutations produced by rotating a single face of the cube?

6.1-3   Which of the following sets of permutations of the cube form a group when combined by the operation "followed by"?
  a)   The set of all permutations which only move cubicles which have a facelet on the Up face.
  b)   The set of all permutations which leave fixed all cubicles which have a facelet on the Up face.
  c)   The set of all permutations which move some cubicle which has a facelet on the Up face.
  d)   The set of permutations obtained by using only F and R turns.

2.   GENERATORS OF A GROUP

Let (S,∗) be a group. Then, given any subset T of S, we can

form a new subset $\langle T \rangle$ consisting of all elements of S which are produced by a finite combination of elements in T and of their inverses. For example, if X, Y, Z are in T then $X*Y^{-1}*Z^{-1}*Y$ is in $\langle T \rangle$. The set $\langle T \rangle$ with the operation $*$ is called the *group generated by T*. That $\langle T \rangle$ is a group is the subject of Exercise 6.2-1. If T consists of a single element X then

$$\langle X \rangle = \{ \cdots, X^{-2}, X^{-1}, I, X, X^2, \cdots \}.$$

If T is a finite set $\{X, Y, \cdots \}$ we also write $\langle T \rangle = \langle X, Y, \cdots \rangle$.

The case in which the group S is *finite*, is of particular interest to students of the cube. Finite groups, of which the permutations on the cube are an example, have some unique properties. Recall from Chapter 4 that if X was the permutation of a process on the cube then, for some positive integer n, we would have the permutation $X^n = I$, where I is the identity permutation. That value of n is called the *order* of X. The same argument extends to an element X from any finite group S. The set $\{X, X^2, X^3, \cdots, X^n, \cdots \}$ can only have a finite number of elements since it is a subset of the finite group, S. Choose n to be the smallest number such that $X^{n+1} = X^k$ with $0 < k \leq n$ — that is, X^{n+1} is the first element in the sequence which equals an earlier element. Then, k must equal 1, because if k > 1 we would have

$$X^n = X^{n+1} * X^{-1} = X^k * X^{-1} = X^{k-1}$$

and for k > 1 X^{k-1} would be in the set $\{X, X^2, \cdots X^n, \cdots \}$. So X^n would be an element in our sequence which equals an earlier one. This is contrary to our assumption that X^{n+1} was the first such element. Hence the assumption of k > 1 is false, that is k must be 1. Notice that the identity and the inverse of each element is in $\langle X \rangle$ since

$$X^n = X^{n+1} * X^{-1} = X * X^{-1} = I$$

and

$$X^{n-1} = X^n * X^{-1} = I * X^{-1} = X^{-1}.$$

From this we see that

$$\langle X \rangle = \{X, X^2, \cdots, X^n\}$$

is the group generated by X. Any finite group that can be generated by a single element is called a *cyclic* group. If n is the number defined above, we say that $\langle X \rangle$ is a cyclic group of *order n*. Note that $X^m = I$ if and only if n divides m evenly. If X is an element of a finite group S then $\langle X \rangle$ is a cyclic group, and the element X^{-1} generates the same cyclic group. However, it is *not* necessarily true for every k that X^k will generate the same cyclic group that X does — see Exercise 6.2-3.

One cyclic group that we encounter immediately on the cube is the set of rotations of a single face. The processes $U, U^2, U^3 = U^{-1}$, and $U^4 = I$, form $\langle U \rangle$, the cyclic group generated by U. An even smaller cyclic group is $\langle U^2 \rangle$. It consists only of U^2 which is its own inverse and $U^4 = I$.

Another interesting cyclic group which is only slightly more complex is the group, $\langle U^2 R^2 \rangle$. Notice that

$$(U^2 R^2)^6 = I$$

is the smallest power of $U^2 R^2$ which produces the identity. Therefore there are six elements in $\langle U^2 R^2 \rangle$, namely $U^2 R^2$, $(U^2 R^2)^2$, $(U^2 R^2)^3$, $(U^2 R^2)^4$, $(U^2 R^2)^5 = (U^2 R^2)^{-1}$, and $(U^2 R^2)^6 = I$.

The number of elements in a group S is called the *order of the group* and is denoted $|S|$. The order of the group generated by any element is called the *order of that element*. Notice that this is consistent with the earlier definition given in Chapter 4 of the order of the permutation of a process on the cube. Determining the order of the entire cube group is an interesting and non-trivial problem. It is the subject of a later section in Chapter 7. Even to determine the order of a small group is not always easy — see Exercise 6.3-1.

EXERCISES: (* indicates harder.)

6.2-1 Show that if T is a non-empty subset of a group S then the set $\langle T \rangle$, of all finite combinations of elements of T and of their inverses, is itself a group.

6.2-2   What conditions must be met by the order of a cyclic group if it contains an element of order 2?

6.2-3   In a cyclic group of order 12,
    a. how many elements have order 1?
    b. how many elements have order 2?
    c. how many elements have order 3?
    d. how many elements have order 4?
    e. how many elements have order 5?
    f.   how many elements have order 6?
    g. how many elements have order 12?

6.2-4   a.   Find a process of order 12 in the cube group.
    b.   In the cyclic group generated by this process find
      i)   a process or order 3
     ii)   a process of order 4
    iii) a process of order 6

6.2-5   Show that every cyclic group, G, is commutative — that is, if x, y are in G then $x * y = y * x$.

6.2-6*   Under what conditions does the cyclic group $\langle X^k \rangle$ have order n for every integer k<n where n is the order of the cyclic group $\langle X \rangle$?

3.   THE TWO-SQUARES GROUP

Our understanding of the entire cube group will be substantially increased by looking at some smaller subgroups generated by a few simple processes. The simplest subgroups are the cyclic subgroups discussed in the previous section. Of the non-cyclic subgroups, the group generated by the 180° rotations of two adjacent faces is one of the most simple and interesting. Thus, $\langle U^2, R^2 \rangle$ is a Two-squares group generated by U^2 and R^2. Can you determine its structure? It happens to have the same structure as the group of rotations and reflections of a regular hexagon.

EXERCISES:

6.3-1 How many permutations are in the Two-squares group, $\langle U^2, R^2 \rangle$?

6.3-2 What is the largest order of any permutation in $\langle U^2, R^2 \rangle$? How many permutations in $\langle U^2, R^2 \rangle$ are there for each order up to and including the largest order?

6.3-3 Show that $\langle U^2, R^2 \rangle$ is not a cyclic group.

6.3-4 Find two processes, X and Y, in $\langle U^2, R^2 \rangle$ which are not the identity and whose permutations are disjoint — that is, such that no piece is moved by both permutations. Show that the group $\langle X, Y \rangle$ is a cyclic group.

4. THE SLICE GROUP

The Slice group is generated by movements of the middle layers, each lying between any two opposite faces. This group and its name were described to the authors by John Conway of Cambridge, England. In our standard notation, this is denoted by $\langle RL^{-1}, FB^{-1}, UD^{-1} \rangle$, however in working with this group it is convenient to introduce some new and more compact notations. We will use the notation

$$R_s = RL^{-1}$$
$$F_s = FB^{-1}$$
$$U_s = UD^{-1}$$
$$L_s = LR^{-1} = R_s^{-1}$$
$$B_s = BF^{-1} = F_s^{-1}$$
$$D_s = DU^{-1} = U_s^{-1}.$$

This is not only more abbreviated but will highlight the use of slice moves in other processes while maintaining the usual centers-fixed coordinates.

It is also enlightening while studying the Slice group to use another coordinate system, the "fixed-corner" coordinate system. This system uses a fixed reference based on a particular corner piece, say the *URF* piece. If the ordinary move F is applied as in Figure 6-1 this orientation system views this move as if the Back layer and the middle layer between Front and Back were turned the other way as in Figure 6-2 leaving the *URF* cubie fixed. Thus to get this orientation system from our "centers-fixed" system, every time a move F, U, or R occurs, it is replaced by $F\mathcal{F}^{-1}$, $U\mathcal{U}^{-1}$, or $R\mathcal{R}^{-1}$ respectively. The moves L, D, and B do not affect the *urf* corner location. Thus they are the same in both the *URF* corner-fixed orientation and the centers-fixed orientation. In the corner-fixed orientation system, we use the notation

$$S_R = RL^{-1}\mathcal{R}^{-1}$$
$$S_F = FB^{-1}\mathcal{F}^{-1}$$
$$S_U = UD^{-1}\mathcal{U}^{-1}$$
$$S_L = R^{-1}L\mathcal{R}$$
$$S_B = F^{-1}B\mathcal{J}$$
$$S_D = U^{-1}D\mathcal{U}.$$

Thus, S_R gives a quarter turn of the middle slice between the Right and Left faces counter-clockwise as viewed through the right face. Similarly, S_U is a quarter turn of the slice between the Up and Down faces counter-clockwise as viewed from above and S_F is a quarter turn of the slice between the Front and Back faces counter-clockwise as viewed through the front. With this notation, the Slice group is

$$\langle S_R, S_F, S_U \rangle.$$

MOVE F, CENTERS-FIXED   MOVE F, *URF* CORNER-FIXED

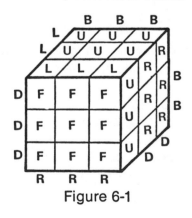

Figure 6-1

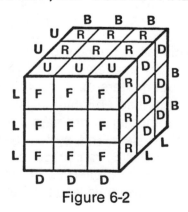

Figure 6-2

Using this notation you will observe that the corner pieces are never moved. Thus the corner pieces can become the fixed reference by which the orientation of the cube can be described. This fixed reference will greatly assist you in understanding the structure of the Slice group.

This group is not too complex and contains some elegant patterns. A person, by sticking only to slice moves, can play with the cube and learn a lot about its movements without getting too lost, even without knowing how to restore the cube.

If you study the permutations which you can achieve in the Slice group, you will see that each face will always show a pattern of facelet colors of the type indicated in Figure 6-3, where a, b, c, and d designate four facelet colors which may or may not be different. Furthermore, the face opposite to this will have each color replaced by its opposite. If we use the notation X' to denote the color of the face opposite the X colored face on a restored cube, then in the slice group the face opposite the pattern of Figure 6-3 will have a', b', c', and d' in place of a, b, c, and d respectively.

SLICE GROUP FACELET PATTERNS

a	c	a
b	d	b
a	c	a

Figure 6-3

EXERCISES:

6.4-1   Show how moves in the Slice group can be used to obtain patterns on all six faces with $a = b' = c' = d$ which is called X or the checkerboard pattern.

6.4-2 Show how to obtain the pattern with $a=b=c\neq d$ on all six faces — this pattern is referred to as spot or sometimes box or even measles.

6.4-3 Show how to obtain the spot pattern with $a=b=c=d'$ on four faces with the solid pattern with $a=b=c=d$ on the other two faces.

6.4-4 Show how to obtain the $+$ pattern with $a=b'=c'=d'$ on four faces and the checkerboard with $a=b'=c'=d$ on the other two faces.

6.4-5 The Slice-squared group is a subgroup of the slice group generated by the squares of the slice moves — that is, $\langle S_U^2, S_R^2, S_F^2\rangle$. It has a particularly simple structure.

    a. Find the permutations of the commutators of the Slice-squared group.

    b. How many elements are in the Slice-squared group?

    c. How many elements are in $\langle S_R^2, S_F^2\rangle$?

5. THE TWO-GENERATOR GROUP

This is an easy group to describe, but it turns out to have a very tricky structure. It is the group generated by the rotations of two adjacent faces, for example, $\langle U, R\rangle$. Even after you have discovered how to restore a scrambled cube using turns of all six faces, it is not easy to find a process which is in $\langle U, R\rangle$ to restore a cube which has been scrambled by only moving the U and R faces. A little experimenting will quickly show you that the edge pieces cannot be flipped in $\langle U, R\rangle$. Other processes which we have already discussed, both conjugates and commutators, are helpful in moving edges and corners to their home locations. Twisting corners with only rotations of the U and R faces may require you to develop a new process.

EXERCISES: (* indicates harder.)

6.5-1 Find a process in $\langle U, R\rangle$ which twists two corners and leaves all other pieces unchanged in their home position.

6.5-2* Describe a method, using only processes in $\langle U, R \rangle$, for restoring a cube which has been scrambled by only rotating the U and R faces.

6. OTHER SUBGROUPS OF THE CUBE

Consider any subset H of the cube group. We say H is closed if for all X and Y in H we have XY in H — that is, H satisfies the Closure Law given in Section 1 of this chapter. If H is closed, it is not hard to show that all the other group criteria hold for H — see the solution to Exercise 6.6-1. Such a set H is called a subgroup of the cube group.

There are many subgroups other than those which we have been discussing in the last three sections. Some of them have been studied in detail by various researchers studying the cube. Here are a few of them.

The Anti-Slice Group. The Anti-slice group, $\langle RL, FB, UD \rangle$, is generated by rotating opposite faces in the same direction — that is, either both clockwise or both counter-clockwise. This rotates them as shown in Figure 6-4 in the manner opposite to that of the Slice group moves. In studying the Anti-slice group, it is again convenient to introduce some special notation. We denote the anti-slice moves by

$$R_a = RL = L_a$$
$$F_a = FB = B_a$$
$$U_a = UD = D_a$$

or to work in the fixed corner coordinates we use

$$A_R = RL\mathscr{R}^{-1} = A_L$$
$$A_F = FB\mathscr{F}^{-1} = A_B$$
$$A_U = UD\mathscr{U}^{-1} = A_D.$$

Figure 6-5 shows how a restored cube is moved by A_R. Like the Slice group, the Anti-slice group also contains several "pretty patterns". Notice also that a slice-squared and an anti-slice-squared are the same. So, the Slice-squared group is a subgroup of the Anti-slice group as well as the Slice group.

THE ANTI-SLICE
MOVE — RL

THE ANTI-SLICE
MOVE — RLR^{-1}

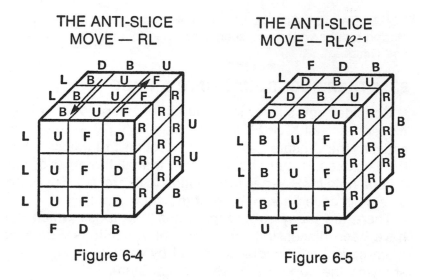

Figure 6-4   Figure 6-5

Multi-Generator Groups. Extending the Two-Generator group to three generators we get either

$$\langle U, R, F \rangle$$

or

$$\langle U, R, D \rangle$$

which have quite different structures. For instance, edge pieces can be flipped in $\langle U, R, F \rangle$ but not in $\langle U, R, D \rangle$. The subgroups generated by the squares of the three generators

$$\langle U^2, R^2, F^2 \rangle$$

and

$$\langle U^2, R^2, D^2 \rangle$$

are also different and interesting.

There are also two different Four-generator groups,

$$\langle U, R, F, D \rangle$$

and

$$\langle U, R, D, L \rangle.$$

The first never moves the *FL* edge piece. The second can never flip the edges. Thus neither is equal to the whole group of the cube.

The Five-generator group $\langle U, R, F, D, L \rangle$ on the other hand, turns out to be equal to the whole group of the cube. To show this, you must find a process using only rotations of the *U, R, F, D,* and *L* faces to form a process which is equivalent to a quarter turn of the *B* face. We leave this as Exercise 6.6-4.

Some other groups of interest are

$$\langle U^2, R^2, F^2, D^2 \rangle$$
$$\langle U^2, R^2, D^2, L^2 \rangle$$
$$\langle U^2, R^2, D^2, L^2, F^2, B^2 \rangle$$
$$\langle U, R^2 \rangle$$

and

$$\langle U, D, R^2, L^2, F^2, B^2 \rangle.$$

The Magic Domino. The last of the subgroups listed above $\langle U, D, R^2, L^2, F^2, B^2 \rangle$ is particularly interesting because of its similarity to the group of the *Magic Domino*. The Magic Domino is a 3 x 3 x 2 version of the cube whose movements are like a cube with the middle layer removed — see Figure 6-6. In the group $\langle U, D, R^2, L^2, F^2, B^2 \rangle$, the edge pieces in the middle layer between the Up face and the Down face never leave that layer. Thus, any algorithm, using only processes in this group for restoring a cube which was scrambled by processes in this group, is an algorithm for restoring a scrambled Magic Domino. If you want to make your cube appear similar to a Magic Domino you can remove or cover the colors on the R, L, F and B faces of the cube and cover the Up face and Down face of the cube with the patterns shown in Figure 6-6.

Another Type Of Subgroup. We have seen one standard method of obtaining a subgroup by considering the sub-

MAGIC DOMINO

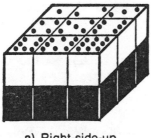 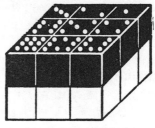

a) Right-side-up   b) Upside-down

Figure 6-6

group generated by a set of elements in a group. Another standard method to obtain a subgroup is to consider the set, H, of elements in a group which preserve some property. If X and Y both preserve some property, so does XY, thus H is a subgroup. For example, the set of processes which leave the *DBR* corner in the *dbr* location is a subgroup. The set of processes which leave all the edges correct is a subgroup. The set of processes which affects only pieces in the Up face is a subgroup, called the U group. Also, the set of processes, which move Up-face pieces only to Up-face locations and arbitrarily permute the other pieces, is a subgroup.

EXERCISES: (* indicates harder)

6.6-1   Given a finite group, (G, *), show that if H is a non-empty subset of G in which the Closure Law is satisfied then H is a subgroup of G.

6.6-2   Find a process in the Anti-slice group which produces
   a.   The "diagonal" pattern shown in Figure 6-7a on four faces and two solid faces.
   b.   The "Z" pattern shown in Figure 6-7b on two faces, the mirror-image of the "Z" pattern, that is with the edge colors interchanged on two faces, and two solid faces.

c. The "2L" pattern shown in Figure 6-7c on six faces.

d. The "+" pattern shown in Figure 6-7d on four faces and two solid faces.

e. The "diagonal" pattern shown in Figure 6-7a on four faces and the "+" pattern shown in Figure 6-7d on two faces.

ANTI-SLICE GROUP FACELET PATTERN

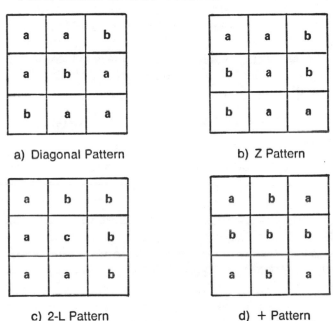

a	a	b
a	b	a
b	a	a

a) Diagonal Pattern

a	a	b
b	a	b
b	a	a

b) Z Pattern

a	b	b
a	c	b
a	a	b

c) 2-L Pattern

a	b	a
b	b	b
a	b	a

d) + Pattern

Figure 6-7

6.6-3   Find a restoration algorithm for $\langle U, R, F \rangle$ — that is, an algorithm using only processes in $\langle U, R, F \rangle$ for restoring a cube scrambled by an unknown process in $\langle U, R, F \rangle$.

6.6-4*   Find a process in $\langle U, R, F, D, L \rangle$ which produces the permutation

(ub, lb, db, rb)
(ulb, ldb, drb, rub),

which is equivalent to the process B.

6.6-5   There are many "pretty patterns" that are not in either the Slice group or the Anti-slice group. For example, find a process which produces the "+" pattern on all six faces.

6.6-6   Find some more "pretty patterns".

6.6-7   Which of the following are subgroups of the cube group?
   a)   The set of permutations which move only the U corners.
   b)   The set of permutations which move some U corners.
   c)   The set of permutations which move *DBR* to *brd*.
   d)   The set of permutations which leave *DBR* in its home location.
   e)   The set of permutations which leave all corners in their home positions and all edges in their home location. (This could be called the flipping group!)

7.   THE SUPERGROUP AND OTHER LARGER GROUPS

One of the first observations made in Chapter 2 about the movements of the cube was that the center pieces of each of the faces have a fixed location relative to one another. Each quarter turn of a face only rotates the center piece of that face in its fixed location. Since that center piece is a solid color, the rotated piece appears indistinguishable from the non-rotated piece. But, what if each center piece were marked so that rotations of the piece were apparent? For example, suppose pictures were printed on each face. Could all the edge and corner pieces be restored to their home positions without restoring the center pieces to their starting orientation? The answer is YES! For example we saw in the solution to Exercise 4.4-1 that

$$(UR)^{105} = I$$

and

$$(UR^{-1})^{63} = I.$$

In the first case, the center of both the Up face and the Right face is rotated a quarter turn clockwise. In the second case, the center of the Up face is rotated a quarter turn counter-clockwise and the center of the Right face is rotated a quarter turn clockwise. However, it is not possible for a process to produce an identity on all the edge and corner pieces while producing a single quarter turn of only one center piece. One center piece can be rotated halfway around without disturbing any other pieces. However, if the edge and corner pieces are left in their starting positions, then the sum of the number of center piece quarter turns must be an even number. The reason for this will be shown in Chapter 7.

By redefining the equivalence of two processes to mean that they not only produce the same permutation on the edge and corner pieces but also produce the same rotations on the centers, we get a much larger set of non-equivalent processes.

These similarly form a group with the same combining operation of "followed by". This is called the *Supergroup* of the cube. The set of all processes which leave the center pieces unrotated forms a subgroup of the Supergroup. This subgroup of the Supergroup permutations has the same structure as the Cube Group — that is, the group of permutations on the cube with the original definition of equality.

The Supergroup is in turn a subgroup of the group of all permutations that could be produced by taking the cube apart and putting it back together with the edge and corner pieces permuted and center pieces rotated. A still larger group can be obtained by identifying all 54 facelets uniquely — for example, by numbering them. Then remove them and rearrange them, noting on each piece not only its new location but also its orientation. I am sure you can think of still larger groups, but I will not even pose that as an exercise.

EXERCISES:

6.7-1   Find a simple process — fewer than 20 moves — which leaves all the edge and corner pieces in place and

   a.   only rotates the Up-face center piece halfway around.

   b.   only rotates the Up-face center piece a quarter turn clockwise and the Right-face center piece a quarter turn counterclockwise.

   c.   only rotates both the Up-face center piece and the Right-face center piece a quarter turn clockwise.

6.7-2   How many processes which are not equivalent in the Supergroup will produce the same permutation of edge and corner pieces — that is, would be equivalent in the Cube Group?

CHAPTER 7

PERMUTATION STRUCTURES AND THE ORDER OF GROUPS

In the last several chapters, you have probably noticed that we have deferred discussion of several points to Chapter 7. As we were writing those chapters, we kept thinking of questions we wanted to discuss or to pose to you, but which required a concept to be developed here. The concept required is the notion of even and odd permutations. Though this is a basic notion and one which seems simple, the concepts which follow from it are certainly less obvious than the notions we have previously introduced. Consequently, we have deferred it to this chapter.

1. PERMUTATIONS ARE ODD OR EVEN

To determine whether a permutation is odd or even, it is decomposed into a succession of pair exchanges — also called swaps, transpositions, interchanges, or, in the nomenclature of Chapter 4, 2-cycles. For example, if a permutation P consisted of the 5-cycle

(ur, fl, fr, fu, fd)

then P could be written as four successive pair exchanges, P_1 followed by P_2 followed by P_3 followed by P_4 where

$$
\begin{array}{ccccc}
 & P_1 & P_2 & P_3 & P_4 \\
ur & \rightarrow fl & \rightarrow fl & \rightarrow fl & \rightarrow fl \\
fl & \rightarrow ur & \rightarrow fr & \rightarrow fr & \rightarrow fr \\
fr & \rightarrow fr & \rightarrow ur & \rightarrow fu & \rightarrow fu \\
fu & \rightarrow fu & \rightarrow fu & \rightarrow ur & \rightarrow fd \\
fd & \rightarrow fd & \rightarrow fd & \rightarrow fd & \rightarrow ur.
\end{array}
$$

First, the pieces in *ur* and *fl* are exchanged. Then the new piece in *ur* — originally in *fl* — is exchanged with the piece in *fr*. Next the new piece in *ur* — originally in *fr* — is exchanged with the piece in *fu*. Finally, the new piece in *ur* — originally in *fu* — is exchanged with the piece in *fd*. Thus, the permutation P is decomposed into pair exchanges,

$$P = P_1 P_2 P_3 P_4 = (ur, fl)(ur, fr)(ur, fu)(ur, fd).$$

Notice that this decomposition is not unique. There are many different ways that P could be decomposed into successive pair exchanges. But, for this particular permutation P, no matter how you decompose P into pair exchanges, the number of pair exchanges needed will always be an even number. More generally, if a permutation consisting of a single cycle is decomposed into successive pair exchanges then the number of pair exchanges needed will always be odd when the length of the cycle is even and will always be even when the length of the cycle is odd. A rigorous proof of this is too elaborate for this book. It can be found in numerous texts or from your teacher. Finally, if a permutation consists of several disjoint cycles, then the number of pair exchanges needed in its decomposition will be odd if the number of even-length cycles is odd. We say that a *permutation is odd* if it decomposes into an odd number of pair exchanges. We say that a *permutation is even* if it decomposes into an even number of pair exchanges. Whether a permutation is odd or even is called the *parity of the permutation*. If P can be written as a product of *m* 2-cycles and Q can be written as a product of *n* 2-cycles, then PQ can be written as a product of *m+n* 2-cycles. Thus parity for the *product* of permutations behaves like the *addition* of odd and even integers.

It all seems trivial enough, doesn't it. Why all the fuss? Patience! Do the exercises and then we will see what follows.

EXERCISES:

7.1-1   Show two decompositions of the permutation *(ub, uf, df)* into two successive pair exchanges and one decomposition into four pair exchanges.

7.1-2   a.   Show that if P_1 and P_2 are both odd permutations then the permutation $P = P_1 P_2$ is an even permutation.
   b.   What is the parity of P if P_1 and P_2 are even?
   c.   What is the parity of P if P_1 is odd and P_2 is even?

7.1-3   Find a process which exchanges only a single pair of corner pieces and a single pair of edge pieces.

7.1-4   Let G be any group of permutations. Show that if G contains any odd permutation then exactly half the elements of G are odd.

2.   PARITY OF PERMUTATIONS ON THE CUBE

To start with, consider the permutation of a single quarter turn of one face. For example, turning the Up-face gives the permutation

(uf, ul, ub, ur)
(ufl, ulb, ubr, urf).

Since this permutation is made up of two disjoint 4-cycles, it is an even permutation. Similarly, a quarter turn of any face produces an even permutation. Hence any process which can be decomposed into a succession of quarter turns of faces will produce an even permutation. But this includes all processes on the cube. That is, all processes on the cube produce even permutations. This now proves what we had guessed all along, namely,

No process on the cube can·exchange a single
pair of pieces while leaving all other pieces in
place.

A single pair exchange is an odd permutation, so it cannot
be produced by any sequence of face turns.

Next we see that a quarter turn of a face produces an odd
permutation of edges and an odd permutation of corners.
Hence any process which decomposes into an odd number
of single-face quarter turns produces an odd permutation of
edges and an odd permutation of corners. Conversely any
permutation which requires an odd permutation of edges
will require an odd number of face quarter turns.

In particular, notice that the identity is an even permutation of edges. Hence any identity on the edge pieces requires an even number of single face quarter turns. Similarly, an identity on corners also requires an even number of
quarter turns. This proves the assertion made in Section 7 of
Chapter 6 that in the Supergroup of the cube it is not possible
to produce a single quarter turn of a center piece while leaving
all edges, corners, and other center pieces in place.

In the restoration process of Chapters 3 and 5, the orientation of five of the six center pieces is fixed as soon as the
four edges of the starting face are restored at the end of
Step 1. It is not until Step 4 that the orientation of the center
piece of the sixth face is decided. The first thing that is
done in Step 4 is to decide where, on the Up face — the sixth
face — to place the first edge piece to be restored on that
face. "Where" in this case means "on which side of the Up-
face center piece". This of course determines the orientation of that center piece and thus on which side of the center piece each of the other Up-face edges must be placed.
Up to this time, not knowing any better, we have chosen this
orientation arbitrarily. As a result, half the time when we
tried to restore the final Up-face edge piece, we found that
we had to move all the previously restored edges one place
clockwise or counter-clockwise around on the Up face. You

may now see why this was necessary, namely to produce an even number of single-face quarter turns.

Furthermore, we can now see how to do better. Before starting Step 4 we can predict which two orientations of the Up face can be used for an identity and which two cannot. To do this, consider the five edge pieces, shown by *, in Figure 7-1, which are to be restored in Step 4. Write down the permutation which is required to restore them to their home positions. If this is an odd permutation, then it requires an odd number of quarter turns of the Up face. Hence, you should start by making that quarter turn of the Up face before determining the home position for the first edge piece to be restored. If the permutation of those five edges is even, then the center piece of the Up face is in an acceptable orientation for placing the first Up face edge in its home position.

FIVE EDGE PIECES TO BE RESTORED IN STEP 4

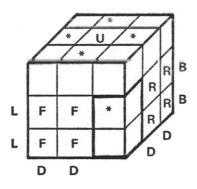

Figure 7-1

EXERCISES: (* indicates harder.)

7.2-1 Two permutations have the same cycle structure if one is a conjugate of the other. After Step 4 of the restoration process has been completed, how many different permutation cycle structures may be needed to describe the possible locations of the corners disregarding twists?

7.2-2 Determine the parity of the permutation of edge pieces produced by any commutator.

7.2-3* Consider the subgroup of the Slice group for which every face has either a "spot" pattern or a solid color, that is, we have $a=b=c$ in the notation of Figure 6-3. The center color, d, may or may not be different. Prove that this subgroup has 12 elements.

7.2-4* Show that every permutation of the cube which is an even permutation of both edges and corners can be produced by a sequence of conjugates of $[F,R]=FRF^{-1}R^{-1}$.

3. THE PARITY OF FLIPS AND TWISTS

We are now ready to consider the assertions made earlier that edge flips come only in pairs and that corner twists come only in pairs or triples — pairs of twists in opposite directions or triples of twists in the same direction. First, we will give you the intuitive argument. Then we will formalize the argument to make it rigorous.

Consider any quarter turn of a single face, say the Up face. This process U produces the permutation

$$urf \rightarrow ufl$$
$$ufl \rightarrow ulb$$
$$ulb \rightarrow ubr$$
$$ubr \rightarrow urf$$
$$uf \rightarrow ul$$
$$ul \rightarrow ub$$
$$ub \rightarrow ur$$
$$ur \rightarrow uf.$$

As each of the four Up-face corners are moved by this quarter turn, the orientation of each one is changed. The sum of these orientation changes is attributable to the move U. We can sum these four orientation changes on a single corner by following one corner as we apply U four times. But U^4 produces no orientation change on that one corner, so we conclude that the sum of the orientation changes is zero.

The same argument shows that the sum of edge orientation changes is zero. Since all processes consist of a sequence of face quarter turns, the orientation changes for any process must add up to zero. It turns out that this argument is fundamentally sound. But, a mathematician would consider it imprecise and would not be sure of its validity until it could be presented rigorously. An example of this imprecision is that no definition is given of the quantity being summed up. What is meant by "the orientation change" of a single corner, say *urf*, when it is moved by U, producing *urf* → *ufl*? The following discussion will use a more specific argument in a more rigorous fashion. The technique used here was first seen by us in draft versions of the forthcoming book *Winning Ways* by J.H. Conway, E.R. Berlekamp, and R.K. Guy where this basic analysis of the Cube Group is attributed to Anne Scott.

Placing the restored cube in a fixed orientation — that is, designating an Up-face color and a Right-face color — we assign to each edge and corner cubicle on the cube a *chief face*. For all cubicles on the Up layer, the Up face will be the chief face, and for all cubicles on the Down layer, the Down face will be the chief face. For the edge cubicles in the middle layer — between the top and bottom — the chief face will be the Right face for those cubicles on the right and the Left face for those on the left. This is shown in Figure 7-2. We also assign a *chief facelet* to each edge and corner cubie in the cube. The chief facelet of each cubie is the facelet which, when the piece is in its home position, matches the chief face of that cubicle.

As pieces are moved around on the cube, their chief facelets never change. We must examine the relation of these chief facelets of cubies to the chief faces of the cubicles they occupy. When the chief facelet of a piece is in the position of the chief face of the cubicle it occupies, it will be called *sane*. Otherwise, it will be called *flipped* if it is an edge piece or called *twisted* if it is a corner piece. A corner piece can be twisted from its sane position in either of two ways,

X INDICATES THE CHIEF FACE FOR EACH CUBICLE

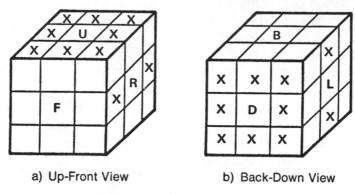

a) Up-Front View   b) Back-Down View

Figure 7-2

clockwise or counter-clockwise. We will show that if we count the number of flips and add up the twists with clockwise counting $+1$ and counter-clockwise counting -1, then the number of flips must be even and the sum of the twists must be a multiple of three.

Notice that in a restored cube, every piece is sane. Hence, the number of flips and the sum of twists are both zero. Again we look at what happens with a quarter turn of a single face. If the Up face or the Down face is turned, no facelet enters or leaves that Up or Down face. The flips or twists of pieces on that layer — as well as for the rest of the cube — remain unchanged. Rotating the Front or Back faces — as in Figure 7-3 — by a quarter turn leaves every flipped edge piece flipped and every sane edge piece sane, because the chief face of each edge cubicle is placed in the chief face of another edge cubicle. Of the four corner pieces on the Front face, two — the pieces in *ufl* and *frd* — get twisted clockwise and two — the pieces in *urf* and *fdl* — get twisted counter-clockwise. Similarly on the Back face, two corners are twisted clockwise and two counter-clockwise. Rotating the Right or Left faces — as in Figure 7-4 — has the same effect on four corners, namely two get twisted

MOVE F   MOVE R

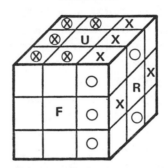

The X indicates the chief face of
each cubicle. The O indicates
the chief facelet of each cubie.

The X indicates the chief face of
each cubicle. The O indicates
the chief facelet of each cubie.

Figure 7-3   Figure 7-4

clockwise and two get twisted counter-clockwise. In these
cases, all four edge pieces are changed. The flipped become sane and the sane become flipped.

From this we can conclude that the number of flips which
we count must be a multiple of four, right? WRONG!!! For
that to be true, you would need to be able to tell the difference between flipping an edge twice and flipping it four
times. But, we can tell the difference between an edge flipped an odd number of times and an edge flipped an even
number of times. Since the number of flips can only be increased four at a time — with each Right or Left face quarter turn, and decreased two at a time — every time the same
edge is flipped twice, we see that the number of flips which
we can count will always be even.

As for corners, every quarter turn adds as many clockwise — +1 — twists as it does counter-clockwise — −1 —
twists. The first temptation is to say that the count of the
number of twists must always be zero. But, three twists in
the same direction on the same corner cannot be distin-
guished from no twist. Similarly, two twists in one direction

looks like one twist in the opposite direction. Thus, although each quarter turn has a zero effect on the sum of twists, three twists are added or subtracted every time a corner is twisted more than once in the same direction. However, the sum of the number of twists must remain a multiple of three.

Solomon W. Golomb has observed an analogy between the subatomic particles called quarks and corners of the cube. Quarks are believed to have charges of $\pm^1/_3$ and never to appear except in combinations where the total charge is integral. Analogously, the corners of the cube can be twisted $\pm^1/_3$ (of 360°) but we can never obtain patterns except where the total twists are integral.

EXERCISES:

7.3-1   What is the smallest number of moves which leaves exactly ten edge pieces "sane" — disregarding corners? A single move can be either a quarter turn or a half turn of a face.

7.3-2   What is the smallest number of moves needed to leave exactly six "sane" corners — disregarding edges?

7.3-3   Show that no corners can be twisted or edges flipped in the Squares group $\langle U^2, R^2, F^2, D^2, L^2, B^2 \rangle$.

7.3-4   Show that no edges can be flipped in the Two-Generator group.

7.3-5   Find a process which produces a totally insane cube which has every piece in its home location — that is, every corner is twisted and every edge is flipped.

4.   THE ORDER OF THE CUBE GROUP

How many different permutations of the cube are there, including different flips and twists of pieces as different permutations, but not distinguishing between different orientations of the cube as a whole? It might be reasonable to leave this as an "exercise for the reader". Perhaps you would like to work it out for yourself before going further. In the following paragraphs, we will go through it to bring out several points.

First, how many permutations of those eight corner locations are possible if we ignore twists? We know — see Exercise 7.4-1 — that the total number of possible permutations of a set of n objects in n locations is n! — n factorial. We have found a process — see Exercise 7.1-3 — which will exchange a single pair of corner locations, ignoring edges, without disturbing any other corners. Using either conjugates of this process or using it with different orientations of the cube, we can exchange any pair of corner locations without disturbing the others. Since every permutation can be decomposed into successive pair exchanges, every permutation of corner locations is possible on the cube. Therefore we have

$$8! = 40,320$$

possible permutations of corners on the cube.

The same argument shows that there are

$$12! = 479,001,600$$

possible permutations of edge locations on the cube. However, we have seen that not all permutations of corner locations can go with all permutations of edge locations. The permutation of corners and edges together must be even, so that even corner permutations must occur with even edge permutations and odd with odd. Can every even corner permutation occur with each even edge permutation? Yes! To see this apply the method above to obtain any even corner permutation. This will produce an even edge permutation which we then need to transform to any desired even edge permutation. This transformation must be an even edge permutation which leaves corners fixed. Any even edge permutation, leaving corners fixed, can be obtained by using either 3-cycles or pairs of 2-cycles of edges which leave corners fixed. These are obtainable as conjugates of either

$$W = LR^{-1}F^2L^{-1}RU^2$$

or

$$V = (F^2R^2)^3.$$

Can every odd corner permutation occur with each odd edge permutation? Yes! We can repeat the above argument and find that we again need to obtain the even edge permutations which leave corners fixed. Alternatively, we simply apply any face turn to transform an odd-odd case into an even-even case, and use the previous argument. Since half the permutations are odd and half are even, we see that the total number of permutations of the cube not counting twists or flips is

$$\frac{8! \ 12!}{2} = 9{,}656{,}672{,}256{,}000.$$

We now consider twists and flips. Any corner can be twisted in three orientations, except for the last one whose orientation is fixed by the other seven. Similarly any of the edges can be flipped in two ways except for the last whose orientation is determined by the first 11. Thus each permutation of the cubies can have

$$\frac{3^8}{3} = 2{,}187$$

corner orientations and

$$\frac{2^{12}}{2} = 2{,}048$$

edge orientations. Thus the total number of cube permutations counting twists and flips is

$$\frac{8! \ 12!}{2} \cdot \frac{3^8}{3} \cdot \frac{2^{12}}{2} = 43{,}252{,}003{,}274{,}489{,}856{,}000$$

$$\approx 4.3 \times 10^{19}$$

This is the order of the entire Cube Group.

EXERCISES:

7.4-1 Show that the number of possible permutations of n objects in n locations is $n! = n \cdot (n-1) \cdots 3 \cdot 2 \cdot 1$. Hint: Notice that $n! = (n-1)! \cdot n$. (The pronunciation for n! is "n factorial").

7.4-2 How many permutations of the Cube Group are there which leave all the edge pieces in their home position?

5. THE ORDER OF THE TWO-GENERATOR GROUP

A similar approach is used to determine the order of the Two-generator group. We first consider edges. We saw in Exercise 7.3-4 that no edges can be flipped in the Two-generator group. It is not hard to find a process which exchanges two edges without disturbing the other edges — it must disturb corners, of course. For example, in $\langle U, R \rangle$ we see that

$$URUR^{-1}URU^2R^{-1}$$

produces the permutation

> *(ur, uf)*
> *(urf, ubr, lbu, luf).*

Using conjugates we can exchange any pair of edges without disturbing the others. Thus we can generate any of the

$$7! = 5040$$

permutations of the seven edge pieces in $\langle U, R \rangle$. But, when it comes to generating permutations of the corners, it is a different story. When you try to show that all permutations of the corners of $\langle U, R \rangle$ can be produced by processes in $\langle U, R \rangle$, you find it can't be done. You find that you can fill any three corner locations with any three corner pieces. But, once you do, then the pieces in the other three corner locations are fixed. We find that the number of possible permutations of the corners of $\langle U, R \rangle$ is only

$$6 \cdot 5 \cdot 4 = 5! = 120.$$

This surprising result is shown in the solution to Exercise 7.5-1. Half of these permutations are odd and half are even. The odd corner permutations go with odd edge permutations and the even go with the even. As before we have a 3-cycle of edges which does not disturb corners. So every odd edge permutation can be produced with each odd corner permutation, and similarly all even edge permutations go with each even corner permutation.

Although edges cannot be flipped, the corners can be twisted — see Exercise 6.5-1 — in pairs. Thus five corners can be twisted arbitrarily and the orientation of the sixth is then fixed. Thus the total number of permutations in the Two-generator group is

$$\frac{7!\cdot 5!}{2}\cdot\frac{3^6}{3}=73,483,200.$$

This may not be close to the order of the Cube Group, but it still is a large number.

EXERCISES: (* indicates harder.)

7.5-1** Show that the number of permutations of corner locations that can be produced in $\langle U, R\rangle$ — disregarding twists — is 120. (This is a very difficult exercise.)

7.5-2* Show that in $\langle U, R\rangle$ the following statements are equivalent:

    a. Disregarding twists, there are not more than 120 possible permutations of corner locations.

    b. It is not possible to have a single 3-cycle of corner pieces leaving the other corners undisturbed — disregarding edge locations and corner twists.

    c. It is not possible to have a single 2-cycle of corner pieces leaving the other corners undisturbed — disregarding edge locations and corner twists.

7.5-3* Find the order of the group $\langle U, R^2\rangle$. You may assume 7.5-1 above.

6. THE ORDERS OF OTHER GROUPS

In Section 6.6 we discussed the Magic Domino, a 3 x 3 x 2 version of the cube whose sides can only be rotated by half-turns whereas the top and bottom layer can be rotated by quarter turns. We observed that the group $G_1 = \langle U, D, R^2, L^2, F^2, B^2 \rangle$ behaved just like the Magic Domino if we ignored the edge pieces in the middle slice between the Up and Down faces. Another group that contains all the permutations of the Magic Domino but does not give all the permutations of edges in the middle slice is $G_2 = \langle U, D, R^2 \rangle$. In this group the effects of L^2, F^2, and B^2 on the U and D faces are produced by

$$L^2 \approx U^2D^2R^2U^2D^2$$
$$F^2 \approx U^{-1}DR^2UD^{-1}$$
$$B^2 \approx UD^{-1}R^2U^{-1}D.$$

Of course, G_2 is a subgroup of G_1. We will see in the next chapter — Exercise 8.3-8 — that it is a special type of subgroup, but for now we are concerned with the order of the groups. We will see in Exercise 7.6-1 that the order of G_1 is

$$(8!)^2 \cdot 12 = 2^{16} \cdot 3^5 \cdot 5^2 \cdot 7^2 = 19,508,428,800$$

and the order of G_2 is

$$(8!)^2 = 2^{14} \cdot 3^4 \cdot 5^2 \cdot 7^2 = 1,625,702,400.$$

The difference reflects the fact that there are 12 even permutations of edges in the middle slice.

So do you think the order of the group of permutations on the Magic Domino is the same as the order of G_2? Yes? Wrong again! One of the permutations counted in G_2 is produced by the process UD^{-1}. But with no middle layer in the Magic Domino to provide a point of reference, this is simply a reorientation of the entire domino and hence doesn't count. Four different permutations in G_2 always turn out to be reorientations of the domino. Thus the order of the group of permutations on the Magic Domino is $(8!)^2/4 = 2^{12} \cdot 3^4 \cdot 5^2 \cdot 7^2 = 406,425,600$.

The order of the Supergroup of the cube is easier to determine. For each permutation of edges and corners, we can turn each of the six center pieces into any of four orientations — rotations — except for the last one whose possible orientations — rotations — are limited to two possibilities when the orientations of the first five are fixed — see Section 7 of Chapter 6. Thus the Supergroup has 2^{11} times as many permutations as does the Cube Group. The order of the Supergroup is then

$$2^{11} \cdot \frac{8!}{2} \cdot \frac{12!}{3} \cdot \frac{3^8}{2} \cdot \frac{2^{12}}{2} = 2^{38} \cdot 3^{14} \cdot 5^3 \cdot 7^2 \cdot 11$$
$$= 88,580,102,706,155,225,088,000$$
$$\approx 8.9 \times 10^{22}$$

The order of the groups obtained by taking the cube apart and putting it back together is still larger — see Exercise 7.6-2.

EXERCISES: (* indicates harder.)

7.6-1 Show that the order of
   a. $G_1 = \langle U, D, R^2, L^2, F^2, B^2 \rangle$ is $(8!)^2 \cdot 12$.
   b. $G_2 = \langle U, D, R^2 \rangle$ is $(8!)^2$.

7.6-2 Find the order of the group obtained by taking a cube apart — *without* removing the facelet colors — and reassembling it:
   a. without regard to rotations of center pieces;
   b. including consideration of center-piece rotations.

7.6-3* Find the order of each of the following groups.
   a. $\langle U^2, R^2 \rangle$
   b. $\langle U, R, D \rangle$
   c. $\langle U^2, R^2, D^2 \rangle$
   d. $\langle U, R, F \rangle$
   e. $\langle U^2, R^2, F^2 \rangle$
   f. $\langle U^2, R^2, F^2, B^2 \rangle$
   g. $\langle U^2, R^2, F^2, B^2, L^2 \rangle$
   h. $\langle U^2, R^2, F^2, B^2, L^2, D^2 \rangle$
   i. $\langle U, R^2, L^2, F^2, B^2 \rangle$

7.6-4   The center, C, of a group G is defined to be the set of all elements of G which commute with every element in G. That is, X is in C if and only if

$$x * g = g * x$$

for every g in G.

   a.   Show that C is a subgroup of G.
   b.   Find the order of the center of the Cube Group — Hint: see Exercise 4.6-1.

CHAPTER 8

ADVANCED RESTORATION METHODS

Students of the cube have devised many methods for restoring the cube. Using the theory of groups and subgroups, we can analyze and compare these methods.

1. NESTED SUBGROUPS

We begin by analyzing the worst case number of moves required by a given method to restore any scrambled cube where a *move* consists of any rotation of a single face. Any quarter or half turn, clockwise or counter-clockwise is a single move. In some contexts, it is preferable to count half turns as two moves, but in this analysis we will not do that.

Analyzing the method of Chapters 3 and 5, we find that the state of the cube at the end of each of the six steps in the method can be described by a subgroup of the cube. For example, at the end of Step 1, the permutation of the cube is in the subgroup G_1 of all the permutations of the cube which leave the Down-face edge locations fixed. At the end of Step 2 the permutation of the cube is in the subgroup G_2 of permutations which leave all Down-face locations fixed except the *drb* corner which was selected as the working corner.

After Step 3 the cube permutation is in the subgroup G_3 of all those permutations which only move pieces on the Up face and in the working corner and working edge locations, *drb* and *rb*. We see that G_3 is a subgroup of G_2, and G_2 is a subgroup of G_1.

Continuing in this manner, after you have completed Step 4 by placing all the edge pieces in their home positions, the permutations that can exist on the five yet-to-be-restored corners form the group G_4 of all permutations of the cube which move only those five corners. Again, G_4 is a subgroup of G_3.

After all the corners are placed in their home locations In Step 5 and only the untwisting processes remain to be done to restore the cube, the possible permutations are the group G_5 of all permutations which only twist, in place, the four Up-face corners and the working corner. Finally, after untwisting the corners we have reduced the permutation of the cube to the identity permutation, I. Thus we have $G_6 = \langle I \rangle$. The given six-step restoration method is characterized by six nested subgroups

$$G \supset G_1 \supset G_2 \supset G_3 \supset G_4 \supset G_5 \supset G_6.$$

For any set S of locations in the cube, there is a corresponding subgroup $G(S)$ in the Cube Group G consisting of all permutations which leave all the locations of S fixed. This follows easily from the discussions in Section 6 of Chapter 6. For any restoration method which gradually restores more and more of the cube, we can define a sequence of sets of pieces of the cube,

$$S_0 \subset S_1 \subset S_2 \subset \cdots \subset S_{n-1} \subset S_n,$$

where S_0 is the empty set, S_n is the set of all pieces of the cube, and S_i is the set of pieces which have been restored after Step i of the restoration method. Then the subgroups

$$G(S_i) \supset G(S_{i+1})$$

form a nested sequence which characterize that restoration method. Thus any restoration method which gradually fixes more and more of the cube can be characterized by a nested sequence of subgroups. But the converse is also true. That is, any nested sequence of subgroups can be used to define a method for restoring the cube. Consider a nested sequence of subgroups

$$G = H_0 \supset H_1 \supset H_2 \supset \cdots \supset H_{n-1} \supset H_n = I.$$

Any method that enables you to reduce every permutation in H_{i-1} to a permutation in H_i, for $i = 1, 2, \ldots, n$, becomes a method for restoring the cube. Furthermore, nothing says that the subgroups H_i need to be of the form $G(S)$ described above. In fact, the subgroup G_5 in the sequence above is not of that form.

An English mathematician, Morwen Thistlethwaite, has developed a method for solving the cube based on a sequence of subgroups, none of which are of the form $G(S)$. Using his method, it is hard for a spectator to see that the cube is close to being restored until the final several twists when suddenly everything falls into its home position. The sequence of subgroups which he uses starting with the entire cube group G are

$$H_0 = \langle L, R, F, B, U, D \rangle = G$$
$$H_1 = \langle L, R, F, B, U^2, D^2 \rangle$$
$$H_2 = \langle L, R, F^2, B^2, U^2, D^2 \rangle$$
$$H_3 = \langle L^2, R^2, F^2, B^2, U^2, D^2 \rangle$$
$$H_4 = \langle I \rangle$$

Thistlethwaite has shown that it is possible to reduce any permutation of the cube to a permutation in H_1 in no more than seven moves. Permutations in H_1 can be reduced to permutations in H_2 in no more than thirteen moves. Then fifteen moves will be sufficient to reduce any permutation in H_2 to a permutation in H_3. Finally he has shown that seventeen moves are all that are ever needed to reduce any permutation in H_3 to the identity permutation in H_4. Thus any scrambled cube can be restored using Thistlethwaite's method in at most 52 moves.

In order to come up with this method and to find the maximum number of moves which might be required to perform each step, Thistlethwaite needed some group theory concepts which we will explain in the next section. Then, using these concepts he was able to make practical use of a computer to help find many of the processes which were needed at each step to reduce any permutation in H_i to some permutation in H_{i+1}. He has not yet given up trying to improve on the number of moves needed for each step. Table 8.1-1 summarizes his results as of this writing, and what improvements he conjectures may come in the future.

Thistlethwaite's Method

Maximum number of moves required	Step 1 $H_0 \rightarrow H_1$	Step 2 $H_1 \rightarrow H_2$	Step 3 $H_2 \rightarrow H_3$	Step 4 $H_3 \rightarrow H_4$	Total
Proven	7	13	15	17	52
Expected to be proven	7	12	14	17	50
Best Possible, this method	7	10	13	15	45

Table 8.1-1

EXERCISES:

8.1-1   Find a set of processes which generate the group of all permutations which leave all the Down-face locations fixed.

8.1-2 Find a set of processes which generate the group of permutations which leave all the Up-face locations and all the Down-face locations fixed.

8.1-3 Show that $\langle U, D, L^2, R^2, F^2, B^2 \rangle$ is a subgroup of the group of all permutation which leave all pieces "sane" — as defined in Section 3 of Chapter 7.

8.1-4 Show that by using a portion of his method for restoring the Cube Group, Thistlethwaite has proved that the Magic Domino can be restored in no more than 32 moves.

8.1-5 Find a set of generators for each of the following groups:
  a) G_1, the set of all permutations which leave the Down face edges fixed.
  b) G_2, the set of all permutations which leave the locations {*dr, dfr, df, dlf, dl, dbl, db*} fixed.
  c) G_3, the set of all permutations which leave all Down-layer and middle layer — between Up and Down — locations fixed except for the *drb* corner and the *rb* edge.
  d) G_4, the set of all permutations of the locations {*drb, ubr, urf, ufl, ulb*} only.
  e) G_5, the set of all permutations which leave all edges fixed, leave the location of all corners unchanged, and only twist corners in the locations {*drb, ubr, urf, ufl, ulb*}.

8.1-6 Is it TRUE or FALSE that the group of all permutations of the cube which are even permutations of both edges and corners is equal to the group of permutations generated by the squares of all processes on the cube?

2. COSETS OF SUBGROUPS

To analyze the number of moves required to reduce a permutation in one group G to a subgroup H of G we make use of the concept of cosets of the subgroup. If (G, ∗) is a group and H is a subgroup of G, then a *coset* of H is defined as follows. For any element x in G, the set Hx of all elements of the form h∗x with h in H is called a *right coset* of H in G. We write for each x in G

$$Hx = \{y : y = h*x \text{ for some } h \text{ in } H\}.$$

The first thing to observe is that the collection of cosets Hx partitions the group G into disjoint — that is, non-overlapping — sets. To say that another way, if Hx and Hy have any element in common, then they are identical — see Exercise 8.2-1. Notice that for any y in G, we have

$$y = I * y$$

and therefore y is in Hy. As a corollary of this and the above mentioned exercise, we see that if y is in Hx then Hx = Hy. This also means that if y is in Hx then

$$x = h * y$$

for some h in H and thus

$$x * y^{-1} = h$$

and

$$y * x^{-1} = h^{-1}$$

are in H.

So what does all this mean about the cube? What it means is this. If H_i and H_{i+1} are two consecutive subgroups in a nested sequence, with

$$H_i \supset H_{i+1}$$

we can determine how many moves are required to reduce a permutation in H_i to a permutation in H_{i+1} by examining the cosets of H_{i+1} in H_i. If X is any permutation in the group H_i, and Y is the "shortest" permutation in the coset $H_{i+1}X$, then applying Y^{-1} to X produces a result which will be in H_{i+1}.

Thus if we are trying to get from a group G to its subgroup H in the shortest number of moves, then we can use the shortest permutations in each of the cosets HX contained in G. Of course, finding the "shortest" permutation in HX may be easier said than done. In many cases it may need the assistance of a computer search.

For a given subgroup H of G, it is natural to ask how many cosets Hx are there? For a finite group G that is easy to answer if we know the order of H and G. First notice that the number of elements in each coset, Hx, is the same as the order of H, because $h_1*x = h_2*x$ if and only if $h_1 = h_2$ — see Exercise 8.2-2. Since the cosets are disjoint then the number of cosets of H in G is equal to the order of G divided by the order of H — that is, $|G|/|H|$. This number is called the *index* of H in G. For example, the subgroup G_1 of the previous section, consisting of all permutations which did not move any edge locations of the Down face, has order

$$|G_1| = \frac{8! \; 8!}{2} \cdot \frac{3^8}{3} \cdot \frac{2^8}{2}.$$

Thus the index of G_1 in the entire Cube Group G is

$$|G| / |G_1| = (\frac{12! \; 8!}{2} \cdot \frac{3^8}{3} \cdot \frac{2^{12}}{2}) / (\frac{8! \; 8!}{2} \cdot \frac{3^3}{3} \cdot \frac{2^8}{2})$$
$$= 11 \cdot 5 \cdot 3^3 \cdot 2^7 = 190,080.$$

For each of the 190,080 cosets G_1X of G_1 in G, there is a "shortest" element Y in G_1X such that the number of moves needed to produce Y is less than or equal to the number of moves required to produce any other permutation in that coset. The maximum over all the cosets of the length of these shortest elements is then the maximum number of moves required to reduce any permutation in G to a permutation in G_1.

The same analysis can be used to determine the number of moves required at each step to reduce any permutation in G_i to a permutation in G_{i+1}. The number of cosets to be examined at each step is the index of G_{i+1} in G_i, namely

$$|G_i| / |G_{i+1}|.$$

This analysis is greatly aided by use of computers.

However, use of a computer does not assist in choosing the sequence of nested subgroups. Furthermore, it is often possible to choose the particular solution used at one step to simplify the reduction problem at the next step. An excellent example of that is given by Step 5 and Step 6 of the restoration methods of Chapters 3 and 5. There, the proper choice in positioning the corners can eliminate the need for untwisting. Of course, you could say "just eliminate G_5 from the sequence of nested subgroups". Two observations come to mind. First, the computer would not do that. Second, how should we account for the fact that if we choose a solution to Step 4 which leaves the corner permutation in a 5-cycle then Step 5 may be made more simple? I don't have the answer. However, the detailed analysis has not been done either, and it may turn out that how we solve Step 4 may *not* simplify Step 5.

It is just such a computer search as we have described, supplemented by some intelligent human analysis, that produced the results in Table 8.1-1 for Thistlethwaite's sequence of nested subgroups. Finding the shortest permutation in each coset does not mean looking at every permutation. It means that all the cosets which contain short permutations are identified by gradually considering longer and longer permutations, until all cosets have been associated with a permutation or the number of unassigned cosets is small enough to make individual analysis practical.

EXERCISES:

8.2-1   Show that the right cosets of a subgroup H of G partition G into disjoint sets.

8.2-2   Show that if G is a finite group and H is a subgroup of G then the order of H divides the order of G evenly. This is known as Lagrange's Theorem, named for the French mathematician who first proved it.

8.2-3   Let G_1 through G_6 be the sequence of subgroups defined in Section 1 of Chapter 8 for restoring the cube.
   a.   Find the index of G_2 in G_1.
   b.   Find the index of G_3 in G_2.
   c.   Find the index of G_4 in G_3.

   d.   Find the index of G_5 in G_4.
   e.   Find the index of G_6 in G_5.
   f.   Find the index of G_3 in G.
   g.   Find the index of G_6 in G_3.

8.2-4   Let H_0 through H_4 be defined as in Section 1 of Chapter 8 for Thistlethwaite's method for restoring the cube. Find the index of H_{i+1} in H_i for $i = 0, 1, 2, 3$.

8.2-5   a)   What is the index of $\langle F^2, R^2 \rangle$ in $\langle F, R \rangle$?
   b)   What is the index of $\langle F^2 R^2 \rangle$ in $\langle F^2, R^2 \rangle$?
   c)   What is the index of the subgroup of even permutations of the cube in the entire Cube Group?
   d)   What is the index of the subgroup of all permutations which leave all corner positions fixed in the entire Cube Group?
   e)   What is the index of the subgroup of all permutations which leave all pieces "sane" — as defined in Section 3 of Chapter 7 — in the subgroup of all permutations which leave all the corners "sane"?

3. NORMAL SUBGROUPS AND ISOMORPHISMS

In the previous section we discussed right cosets of a subgroup H in a group G. We can similarly define a left coset of H for any x in G as

$$xH = \{y : y = xh \text{ for some } h \text{ in } H\}.$$

The left cosets similarly partition the group G into disjoint subsets. However in general, the left cosets of H need not be the same as the right cosets. For example, if H equals the four element cyclic group consisting of the four rotations of the Up face, that is,

$$H = \langle U \rangle = \{I, U, U^2, U^3\}$$

then we have

$$FH = \{F, FU, FU^2, FU^3\}$$

and

$$HF = \{F, UF, U^2F, U^3F\}.$$

We can see that

$$FH \neq HF.$$

There are some subgroups however for which the left cosets and the right cosets are the same. That is, we have

$$xH = Hx$$

for all x in G. When this is true, the subgroup H is called a *normal* subgroup of G. Sometimes such subgroups are called invariant or self-conjugate — as suggested by Exercise 8.3-1. Normal subgroups are of particular interest when we are trying to understand the structure of the group. Can you think of any examples in the entire Cube Group G of a normal subgroup? There are several.

The easiest example is derived from observing that any subgroup of index two must be normal, since the two cosets, either right or left, must be the subgroup itself and the complement of the subgroup in the group. The subgroup of all permutations which are even permutations of both corners and edges is a subgroup of index two in the Cube Group G. We will use A to denote this subgroup.

Two other examples of normal subgroups are, first, the subgroup of all permutations which leave all corner positions fixed and, second, the subgroup of all permutations which leave all edge positions fixed. The first, which we will denote by A_e, is the same as all even permutations which move only edges. The second, which we will denote by A_c is the same as all even permutations which move only corners. Notice that A_e and A_c are both normal subgroups of A as well as of G.

The cosets of a normal subgroup have a most interesting property. There is a natural way to define an operation for combining any two cosets, namely

$$(xH) \ast (yH) = (x \ast y)H.$$

Furthermore, when the cosets of a normal subgroup are combined in this way, they form a group — see Exercise

8.3-2. We use G/H to denote this group and it is called the *factor group* of the normal subgroup H in G. It is also sometimes called the *quotient group* of G by H.

What is the structure of the factor group of A in G? It has only two elements and all groups with only two elements have a similar structure. It must be the cyclic group generated by the element which is not the identity. In this case the identity is the coset consisting of the group A itself. The other element is the coset consisting of those permutations which are odd permutations of both corners and edges. It is easy to see that combining any two of these permutations gives a permutation in A.

We now turn to the structure of the factor groups of A and A_c respectively. A little reflection will show you that the factor group of A_e in G has the structure of the group of all permutations of the corners of the cube, where two permutations are considered the same or equal if they only differ on edge pieces. Similarly the factor group of A_c in G has the structure of the group of permutations on the edges of the cube, where the corners are ignored.

We have previously indicated that different groups may have the same structure. Now we will give you the formal definition of what we mean when we say one group has *the same structure* as another. We call two such groups isomorphic. Two groups (G, $*$) and (G′, $\star$) are said to be *isomorphic* if there exists a one-to-one mapping f of G onto G′, that is

$$x \rightarrow f(x)$$

such that

$$f(x*y) = f(x) \star f(y)$$

for every x and y in G. Such a mapping is sometimes called a *product-preserving* mapping. The *one-to-one and onto* properties of f assure that the sets G and G′ have the same size. The *product-preserving* property assures that the operations $*$ and $\star$ have the same structure.

The simplest examples on the cube of two groups that are isomorphic are groups that are different only because of orientation. This gives us another and more rigorous view of the concept of orientation. The group generated by rotating the Up face is not the same exactly as the group generated by rotating the Front face. Rotating the Up face permutes the Up face pieces while rotating the Front face rotates the Front face pieces. But that they both have the same structure becomes obvious when we reorient the cube to place the whole Front face on top in place of the Up face. We can then rotate the now-Up face just as we could before the reorientation. If we then move the now-Up face back to the front we have produced a Front-face rotation. This procedure suggests the mapping which defines an isomorphism between $\langle U \rangle$ and $\langle F \rangle$. Define the mapping f for each X in $\langle U \rangle$ to be

$$f(X) = RXR^{-1}$$

which is in $\langle F \rangle$. Then we have

$$f(XY) = RXYR^{-1} = RXR^{-1}RYR^{-1} = f(X)f(Y),$$

so f is a product-preserving mapping. Since R is itself a permutation of the cube pieces, it is included along with X and Y in the group of all permutations of all cube pieces. In this group we see that

$$RXR^{-1} = RYR^{-1}$$

if and only if

$$X = Y.$$

Also for each Z in $\langle F \rangle$ we have $X = R^{-1}ZR$ in $\langle U \rangle$ with

$$f(X) = Z.$$

Thus f is seen to be a one-to-one and onto mapping. Hence f is an isomorphism between $\langle U \rangle$ and $\langle F \rangle$.

A similar argument can show that any other two groups which only differ by their orientation are isomorphic. However, the concept of isomorphism applies to much more than different orientations. For example, notice that all two-element groups are isomorphic. In particular, notice that the factor group G/A is isomorphic to any two-element cyclic group.

Notice next that A_e and A_c are both normal subgroups of A. We observe — see Exercise 8.3-3 — that the factor group A/A_e is isomorphic to A_c and similarly the factor group A/A_c is isomorphic to A_e. Furthermore any element in A can be written uniquely as the combination of an element in A_e and an element in A_c.

Can we find any normal subgroups of A_e and A_c? The answer is YES! The set A_f of all permutations in A_e which leave the locations of all edge pieces fixed but may flip any even number of them is a normal subgroup of A_e. The set A_t of all permutations in A_c which leave all the corner locations fixed except for twisting some of them is a normal subgroup of A_c. The factor group A_e/A_f is isomorphic to a subgroup A_{se} of A_e consisting of all even permutations of edges which leave all the edges "sane" — see Section 3 of Chapter 7. The factor group A_c/A_t is isomorphic to a subgroup A_{sc} of A_c consisting of all even permutations of corners which leave all corners "sane". Notice that A_{se} and A_{sc} are *not* normal subgroups of A_e and A_c respectively. However A_{se} and A_{sc} are isomorphic to well known groups. These two isomorphisms enable us to complete our analysis of the factorization of G into normal subgroups by showing that there are no more such subgroups.

It is not hard to see that the group of all permutations of one set of n objects is isomorphic to the group of permutations of any other set of n objects — if you disregard the orientations of the objects in their locations. This group of all permutations of n objects is called the n-element *symmetric* group, denoted by S_n, and as we saw in Exercise 7.4-1 this group has order n!. The subgroup of all even permutations of S_n is a normal subgroup of index two. It is called the n-element *alternating* group and is denoted by A_n. The group

A_{se} is isomorphic to A_{12} and the group A_{sc} is isomorphic to A_8.

It is a famous result of group theory that the groups A_n, of even permutations of n objects, have no non-trivial normal subgroups for $n \geq 5$. Of course, the identity I and the whole group are trivial normal subgroups of any group. In about 1821, the 19 year old Niels Henrik Abel showed that there is no method for solving polynomial equations of the fifth or higher degrees by means of a finite sequence of algebraic processes — that is, addition, subtraction, multiplication, division, and extraction of roots. In 1831, the 20 year old Evoriste Galois clarified Abel's ideas and shows that Abel's result is a consequence of A_n having no non-trivial normal subgroups for $n \geq 5$.

A group with no non-trivial normal subgroups is called *simple*. It turns out that simple groups are the basic building blocks of group theory just as the primes are the building blocks of number theory. The determination of the finite simple groups has been a major goal of mathematics for the last twenty years and was recently completed in 1980.

EXERCISES: (* indicates harder.)

8.3-1   Show that H is a normal subgroup of G if and only if $xHx^{-1} =$ H for every x in G.

8.3-2   Show that if H is a normal subgroup of G using the operation *, then the set of left cosets of H form a group G/H, using the combining operation
$(xH)*(yH) = \{z : z = (x*h_1)*(y*h_2)$ for some h_1 and h_2 in H $\}$.
Hint: this is actually a fairly easy exercise except for one detail. One must check that $xH*yH = (x*y)H$ really defines an operation on cosets. That is, is it true that $(x*y)H$ is a left coset of H? This is where you need the fact that H is normal.

8.3-3   Show that A_e is a normal subgroup of G — Note: A similar argument shows that A_c is also a normal subgroup of G.

8.3-4   Show that the factor group A/A_e is isomorphic to A_c.

8.3-5   Other than the identity and the entire group, find a normal subgroup of

a) $\langle F^2, R^2 \rangle$
b) $\langle FR \rangle$
c) The Slice Group.

8.3-6 A group G′ is said to be a homomorphic image of a group G if there is a mapping f — usually not 1-1 —

$$x \rightarrow f(x)$$

from G onto G′ such that

$$f(x) \cdot f(y) = f(x \cdot y)$$

for all x and y in G. Show that G/A_c is a homomorphic image of the entire cube group, G — Note: For any normal subgroup H of group G, the factor group G/H is a homomorphic image of the group G.

8.3-7 Find a mapping by which the group of permutations of the Magic Domino is a homomorphic image of

$$H_2 = \langle U, D, L^2, R^2, F^2, B^2 \rangle.$$

8.3-8 Which of the following statements are true and which are false?

a) The set S of processes in H_2, which in Exercise 8.3-7 map into the identity permutation of the Magic Domino, form a normal subgroup of H_2.
b) The even permutations S_e of the edges in the middle layer between the Up and Down faces is a normal subgroup of H_2.
c) The set S_e is a normal subgroup of G.
d) The set S_e is a subgroup of S and the index of S_e in S is four.

8.3-9 a) Is $\langle R \rangle$ a normal subgroup of $\langle U, R \rangle$?
b) Is the Cube Group G a normal subgroup of $\langle U, R, U, D, R, L, F, B \rangle$?
c) Is $\langle U \rangle$ a normal subgroup of $\langle U, U \rangle$? What is the index of $\langle U \rangle$ in $\langle U, U \rangle$?

8.3-10 Which of the following statements are true and which are false?

a) The groups, A_f and A_t, are commutative groups.
b)* The group of permutations of the Magic Domino is a homomorphic image of the entire Cube Group.
c) The Cube Group is a normal subgroup of the Supergroup of the cube.
d) The Cube Group is a normal subgroup of the group of all permutations produced by taking the cube apart and putting it back together with the pieces rearranged.

CHAPTER 9

EPILOGUE

At this point one might say, "Isn't it natural to restore the cube using some sequence of the groups A, A_e, A_f, A_{se}, A_c, A_t, or A_{sc}". As we have seen in Chapter 8, most methods follow this approach with minor variations. At the very least people tend to work on corners and edges separately. In analyzing how many moves are required to restore the cube, very little is considered regarding what might be done while restoring A_e to simplify the restoration of A_c or visa-versa. Rather each step is considered independently of the others and evaluated assuming a worst case input and an arbitrary output.

The best known restoration method to date is that of Thistlethwaite given in Section 1 of Chapter 8. No one knows how many moves would be needed for "God's Algorithm" assuming he always used the fewest moves required to restore the cube. It has been proven that some patterns must exist that require at least seventeen moves to restore but no one knows what those patterns may be. Experienced group theorists have conjectured that the smallest number of moves which would be sufficient to restore any scrambled pattern — that is, the number of moves required for "God's Algorithm" — is probably in the low twenties. We make a further conjecture. We conjecture that as long as one considers corner permutations and edge permutations separately, it will not be possible to analyze a method to show that fewer than forty moves are required.

To do better, to get closer to "God's Algorithm", we must recognize for any permutation the coset into which it falls for several subgroups at once. And then — as Douglas Hofstadter might point out, "much in the manner of a Bach canon or an Escher print" — we must find an inverse permutation which weaves together the inverses for the several cosets to simultaneously reach the identity of each factor group, and thus obtain the identity for the entire Cube Group.

APPENDIX A

A SMALL CATALOGUE OF PROCESSES

Many processes which produce particularly simple or interesting permutations have been discovered, rediscovered, and rediscovered again by many "cubemeisters" over the last several years. Some of these are catalogued below. Processes from certain subgroups of interest are listed together.

We have tried to attribute these to their first discoverer as best we know it, and apologize to anyone whom we have slighted in this endeavor. In any event, most of these processes have been discovered by many people. Processes which have come from several sources are attributed to "well-known". We abbreviate the most common sources as follows:

BCS — Benson, Conway, and Seal
BCG — Berlekamp, Conway, and Guy
AHF — Alexander H. Frey, Jr.
3DJ — 3-D Jackson
GK — Gerzson Kéri
KO — Dame Kathleen Ollerenshaw
DBS — David Singmaster
MBT — Morwen Thistlethwaite

159

MBTC — Morwen Thistlethwaite's computer
WK — Well-known.

Following each source we indicate the number of moves in the process. Thus an attribution such as (WK — 10) indicates that the process is well known and takes ten moves.

We first give the permutation and then the sequence of moves. Processes for the inverses of given permutations are not given. Also permutations which are obtained from given processes by re-orienting the cube are not usually given.

The primary subgroups which we have used for this catalogue are two subgroups which have been extensively studied, the subgroup of permutations which move only pieces on a single face — we have chosen the Up face — and the subgroup $\langle U, R \rangle$ of permutations produced by only rotating two adjacent faces. These have been further subdivided into the subgroups, P_C, of permutations which move only corners, and the subgroup, P_E, of permutations which move only edges. Other processes of interest are catalogued under "Other Processes".

Up-Face Corner Permutations.

(i) One 3-Cycle. There are nine possible cases. All others can be obtained from inverses and/or reorienting the cube.

(ulb, ufl, ubr)—B²L²BRBWBR⁻¹B	(MBT — 9)
—LF⁻¹LB²L⁻¹FLB²L²	(GK — 9)
(ulb, ufl, bru) —LU²LDL⁻¹U²LD⁻¹L²	(AHF — 9)
(ulb, ufl, rub) —R⁻¹FRB⁻¹R⁻¹F2¹RB	(3DJ — 8)
—BLFL⁻¹B⁻¹LF⁻¹L⁻¹	(KO — 8)
(ulb, flu, ubr) —LFR⁻¹F⁻¹L⁻¹FRF⁻¹	(3DJ, GK — 8)
—BLFL⁻¹B⁻¹LF⁻¹L	(KO, MBT — 8)
(ulb, flu, bru) —R²F²R⁻¹B²RF²R⁻¹B²R⁻¹	(KO, MBT — 9)
(ulb, flu, rub) —URU⁻¹L⁻¹UR⁻¹U⁻¹L	(DBS, WK — 8)
—BU⁻¹F⁻¹UB⁻¹U⁻¹FU	(DBS — 8)
(ulb, luf, ubr) —B²D⁻¹BU²B⁻¹DBU²B	(GK — 9)
(ulb, luf, bru) —FL⁻¹B²L⁻¹F²LB²L⁻¹F²L²F⁻¹	(DBS — 11)
—BD⁻¹B²D⁻¹F²DB²D⁻¹F²D²B⁻¹	(KO — 11)
(ulb, luf, rub) —BRDR⁻¹U²RD⁻¹R⁻¹U²B⁻¹	(AHF — 10)

(ii) Two 2-Cycles. The easiest known processes for producing many of these come from using two of the 3-cycles given above.

Those are the cases for which no process is listed. It is surprising to find that so many are produced by inverses and reorientation of others. All of the 54 possible cases can be produced by inverses and/or reorientation of the 18 listed below.

(ulb, urf)(ufl, ubr) $—RLU^2R^{-1}L^{-1}F^{-1}B^{-1}U^2FBU^2$

(N.J. Hammond, BCS — 11)

$—R^2F^2B^2L^2DR^2F^2B^2L^2U$ (R. Walker — 10)

(ulb, fur)(ufl, bru) $—L^{-1}FD^2F^{-1}LUL^{-1}FD^2F^{-1}LU^{-1}$ (AHF — 12)

(ulb, rfu)(ufl, ubr) $—L^{-1}U^{-1}LU^{-1}L^{-1}ULU^{-1}L^{-1}ULU^{-1}L^{-1}U^2L$

(AHF — 15)

(ulb, urf)$_+$(ubr, flu)$_-$ $—B^{-1}U^{-1}BU^{-1}BUB^2UB^2U^2B^{-1}U^2$ (MBT — 12)

(ulb, fur)$_+$(ubr, flu)$_-$ $—R^{-1}U^{-1}RU^{-1}R^{-1}U^2RBU^2B^{-1}U^{-1}BU^{-1}B^{-1}U^2$

(AHF — 15)

(ulb, rfu)$_+$(ubr, luf)$_-$ $—F^{-1}L^{-1}FR^{-1}F^{-1}LFR^{-1}D^{-1}RU^2R^{-1}DRU^2R^{-1}$

(AHF — 16)

(ulb, ubr)(urf, ufl) $—(RU^{-1}LD^2L^{-1}UR^{-1}U^2)^2$ (AHF — 16)

(ulb, bru)(urf, flu) $—B(LUL^{-1}U^{-1})^3B^{-1}$ (DBS — 14)

(ulb, rub)(urf, luf) $—(RU^{-1}F^{-1}D^2FUR^{-1}U^2)^2$ (AHF — 16)

(ulb, ubr)(urf, flu)   —

(ulb, ubr)(urf, luf) $—FU^2BD^2B^{-1}U^2F^{-1}UFU^{-1}BD^2B^{-1}UF^{-1}U^{-1}$

(AHF — 16)

(ulb, bru)(urf, luf)   —

(ulb, ubr)$_+$(urf, luf)$_-$ $—R^{-1}F^{-1}RL^2B^{-1}R^{-1}BL^2FB^{-1}RB$ (MBT — 12)

(ulb, bru)$_+$(urf, ufl)$_-$ —

(ulb, rub)$_+$(urf, flu)$_-$ —

(ulb, ubr)$_+$(urf, ufl)$_-$ —

(ulb, ubr)$_+$(urf, flu)$_-$ —

(ulb, bru)$_+$(urf, flu)$_-$ —

(iii) Twists. Reorientations and inverses reduce the number of possible cases here to five.

(ulb)$_+$(ubr)$_-$ $—L(U^2LB^{-1}D^2BL^{-1})^2L^{-1}$

(E. Rubik & WK — 13)

(ulb)$_+$(urf)$_-$ $—(U^2BR^{-1}D^2RB^{-1})^2$ (E. Rubik & WK — 12)

(ulb)$_+$(ubr)$_+$(urf)$_+$ $—U^2LF^{-1}L^2FLF^{-1}L^2FU^2BLB^{-1}L^{-1}$

(MBT — 14)

(ulb)$_+$(ubr)$_+$(urf)$_-$(ufl)$_-$ $—L^{-1}FD^2LF^2D^{-1}FU^2F^{-1}DF^2L^{-1}D^2F^{-1}LU^2$

(MBT — 16)

(ulb)$_+$(ubr)$_-$(urf)$_+$(ufl)$_-$ $—L(FU^{-1}RUR^{-1}UF^{-1}U^{-1})^2L^{-1}$(BCS — 18)

Up-Face Edge Permutations

(i) One 3-Cycle. Reorientations and inverses reduce these to four cases.

(ub, uf, ur)—$R^2U^{-1}FB^{-1}R^2F^{-1}BU^{-1}R^2$		(MBT & WK — 9)
(ub, uf, ru) —$RD^2L^2BL^2D^2R^2FR$		(R. Walker — 9)
(ub, fu, ur) —$L^{-1}B^{-1}R^{-1}URBLFU^{-1}F^{-1}$		(MBT — 10)
—$L^{-1}RUR^{-1}U^{-1}LR^{-1}FRF^{-1}$		(O. Pretzel — 10)
(ub, fu, ru) —$B^{-1}U^{-1}BLFRUR^{-1}F^{-1}L^{-1}$		(MBT — 10)

(ii) Two 2-Cycles. Reorientations and inverses reduce these to nine cases. In many cases, nice solutions are not known.

(ub, uf)(ur, ul)	—$RLU^2R^{-1}L^{-1}F^{-1}B^{-1}U^2FB$	(WK — 10)
(ub, uf)(ur, lu)	—	
(ub, fu)(ur, lu)	—	
(ub, uf)$_+$(ur, lu)$_+$	—$F^{-1}U^{-1}L^{-1}U^2LFURU^2R^{-1}$	(3DJ — 10)
(ub, ur)(uf, ul)	—$F^2B^2D^{-1}L^2F^2B^2R^2F^2B^2DF^2B^2$	
		(R. Walker & 3DJ — 12)
	—$F[U, R]F^{-1}B[L, U]B^{-1}$	(DBS — 12)
(ub, ru)(uf, ul)	—$RBUB^{-1}U^{-1}R^2F^{-1}U^{-1}FUR$	
		(D.E. Taylor & 3DJ — 11)
(ub, ru)(uf, lu)	—$FBR(U^2B^2)^3R^{-1}F^{-1}B^{-1}$	(DBS — 12)
(ub, ur)$_+$(uf, ul)$_+$	—	
(ub, ur)$_+$(uf, lu)$_+$	—	

(iii) Flips. There are only three different cases.

(ub)$_+$(ur)$_+$	—$B^{-1}U^2B^2UB^{-1}U^{-1}B^{-1}U^2FRBR^{-1}F^{-1}$	
		(MBTC — 13)
(ub)$_+$(uf)$_+$	—$U^{-1}FR^{-1}UF^{-1}RL^{-1}UB^{-1}RU^{-1}BR^{-1}L$	
		(F. Barnes & BCS — 14)
(ub)$_+$(ur)$_+$(uf)$_+$(ul)$_+$	—$R^2B^2R^2U^2RL^{-1}BLR^{-1}U^2R^2B^2R^2U$	
		(MBT — 14)

Other Short Up-Face Processes:

(ulb, ubr)(ub, ur)— $F^{-1}UBU^{-1}FU^2B^{-1}UBU^2B^{-1}$	(3DJ & MBT — 11)
(ulb, rub)(ub, ul)— $R^{-1}URU^2R^{-1}L^{-1}URU^{-1}LU^2$	(J. Trapp — 11)
(ulb, rfu)(ub, ul)— $FRUR^{-1}F^2LFL^2ULU^{-1}$	(GK — 11)
(ulb, urf)_(ubr)$_+$(ub)$_+$(ur, uf)$_+$ — $URBUU^{-1}B^{-1}R^{-1}$	
	(F. Barnes & BCS — 7)
(ulb, ubr)$_+$(urf, luf)_(ub, ur, fu) — $FURU^{-1}R^{-1}F^{-1}$	(DBS — 6)
(ulb, ubr)$_+$(urf, luf)_(ub, lu, fu) — $BLUL^{-1}U^{-1}B^{-1}$	(DBS — 6)
(ulb, urf)$_+$(ubr, flu)_(ub, uf, ur) — $FUF^{-1}UFU^2F^{-1}$	(J. Trapp — 7)

(ulb, flu, ubr)(ur)₊(uf)₊ — R⁻¹U²R²UR⁻¹U⁻¹R⁻¹U²FRF⁻¹

$\hfill$ (MBTC — 13)

(ulb)₊(ubr)_(ub)₊(ur)₊ — B²R²FBR⁻¹B⁻¹RF⁻¹R²B²URU⁻¹R⁻¹

$\hfill$ (MBT — 14)

(ulb)₊(ubr)₊(urf)₊(ub, ur, uf) — U²LUL⁻¹ULU²L⁻¹   (BCS & KO — 8)

(ulb)₊(ubr)₊(urf)₊(ub, ur, ul) — U²B⁻¹U²BUB⁻¹UB   (K. Fried — 8)

(ulb)₊(ubr)₊(urf)₊(ub)₊(ur)₊ — B⁻¹U⁻¹B²L⁻¹B⁻¹L²U⁻¹L⁻¹U²

$\hfill$ (D. Benson — 9)

(ulb)₊(ubr)₊(urf)_(ufl)_(ub, ur)(uf, ul) —

$\qquad\qquad$ FURU⁻¹R⁻¹F⁻¹BLUL⁻¹U⁻¹B⁻¹   (MBT — 12)

(ulb)₊(ubr)₊(urf)_(ufl)_(ub, fu, ur) — F(URU⁻¹R⁻¹)²F⁻¹   (DBS — 10)

(ulb)₊(ubr)₊(urf)_(ufl)_(ub, fu, lu) — B(LUL⁻¹U⁻¹)²B⁻¹   (DBS — 10)

*(ulb)₊(ubr)_(urf)₊(ufl)_* — (FUF⁻¹UFU²F⁻¹)²   (J. Trapp — 11)

Corner Processes in ⟨U, R⟩.

(i) Single 3-Cycle. A single 3-cycle of corners is not possible — see Exercises 7.5-1 & 2.

(ii) Two 2-Cycles.
(ulb, bru)(urf, dfr)   —(URU⁻¹R⁻¹)³   (DBS — 12)
(ulb, drb)(ubr, rfu)   —(UR⁻¹U⁻¹R)³   (DBS — 12)
*(ulb, fur)₊(ubr, luf)_* —U²RU²R²U⁻¹R²U⁻¹R⁻¹UR⁻¹UR   (MBT — 12)
*(ulb, fur)₊(ubr, flu)_* —R⁻¹U⁻¹RU⁻¹R⁻¹U²RURU²R⁻¹U⁻¹RU⁻¹R⁻¹U

$\hfill$ (AHF — 16)

(iii) Twists.
*(ulb)₊(ufl)_* — RUR⁻¹URU²R⁻¹U²R⁻¹U⁻¹RU⁻¹R⁻¹U²RU² (AHF — 16)

Edge Processes in ⟨U, R⟩.

(i) One 3-Cycle.
(uf, uf, ul) — RUR⁻¹URU²R⁻¹U⁻¹R⁻¹U⁻¹RU⁻¹R⁻¹U²RU   (AHF — 16)

(ii) Two 2-Cycles.
(ub, uf)(rb, rf) — (U²R²)³   (DBS — 6)

(iii) Flips. No flips of edges are possible — see Exercise 7.3-4.

Other Short Processes in ⟨U, R⟩.

(ulb, ubr)₊(urf, frd)_(ub, ur, fr) — URU⁻¹R⁻¹   (DBS — 4)
(ulb)₊(ubr)₊(urf)_(frd)_(ub, fr, ur) — (URU⁻¹R⁻¹)²   (DBS — 8)
(ulb)_(ubr)₊(urf)_(ufl)₊(rf, lu, fu)(rb, ru, bu) — U⁻¹R²U²R⁻¹U²R²

$\hfill$ (? — 6)

Other Short Corner Processes.

(urf, ulb)(dlf, drb) — $(R^2U^2F^2)^6$ (A.J. Adamyck & WK — 18)
(ufl, rub, frd) — $LD^{-1}B^2DFD^{-1}B^2DF^{-1}L^{-1}$ (AHF — 10)
(ufl, ubr, dfr) — $D^{-1}F^{-1}R^{-1}D^2RFU^2F^{-1}R^{-1}D^2RFU^2D$ (AHF — 14)
(ulb)$_+$(ubr)$_-$(urf)$_-$(ufl)$_-$(dbl)$_-$(drb)$_+$(dfr)$_+$ (dlf)$_+$ — $(LR^2F^2B^{-1})^4$
                                                (MBT — 16)

Other Short Edge Processes.

(ub, db)(uf, df)   — $F^2R^2L^2B^2R^2L^2$ (DBS & WK — 6)
(uf, df)$_+$(ur, dr)$_+$ — $FUF^2U^{-1}R^{-1}F^{-1}U^{-1}R^2UR$ (MBTC — 10)
(ub, df)(uf, db)   — $RL^{-1}U^2D^2R^{-1}LF^2B^2$ (DBS — 8)
(ub, uf, df)   — $F^2RL^{-1}U^2R^{-1}L$ (T. Varga & K. Fried — 6)
(uf, ru, fr)   — $RU^{-1}R^{-1}F^{-1}L^{-1}B^{-1}U^{-1}BLFRU^2R^{-1}$ (GK — 13)
(ul, rb, fd)   — $U^2DFDBR^{-1}B^{-1}D^{-1}F^{-1}U^{-1}RU^{-1}D^{-1}$ (GK — 13)
(ur)$_+$(dr)$_+$(ul)$_+$(dl)$_+$ — $F^2B^2LF^2UD^{-1}R^2BR^2L^2F^{-1}L^2U^{-1}DB^2R^{-1}$
                                                (MBT — 16)

Many other processes are given in the exercises. In particular, Supergroup processes are in the solution to Exercise 6.7-1 and Pretty Patterns are in the solutions to exercises at the end of Section 4 and Section 6 of Chapter 6.

APPENDIX B

SOLUTIONS TO EXERCISES

CHAPTER 2
Section 1.

2.1-1   a) 24   b) 24   c) 6

2.1-2   Five

2.1-3   The corner can only be placed diagonally opposite to its home position — for example, the *UFL* piece can only be placed in the *drb* location.

Section 2.

2.2-1   Four

2.2-2   a) True   b) True   c) False

2.2-3   Twenty-four

Section 3.

2.3-1   Edges; *lb, ld, bd.*
          Corners; *ldb.*

2.3-2   a) F   b) FDR   c) FRU

2.3-3   See Figure C-1.

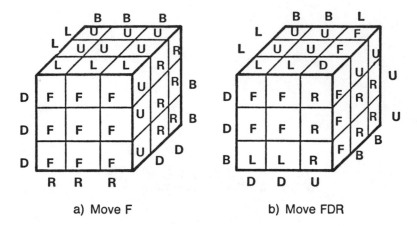

a) Move F   b) Move FDR

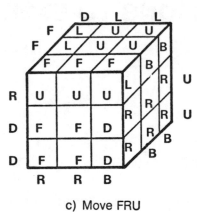

c) Move FRU

Figure C-1

2.3-4   a) *UFL, ULB, UBR, URF*
        b) *df, dl, db, dr*
        c) *FL, LB, BR, RF*
        d) *ufl, ulb, ubr, urf, frd, fdl, drb*
        e) *UL, UB, UR, UF, RF, DF, LF*

2.3-5 $\mathcal{L} = \mathcal{F}\,\mathcal{U}\,\mathcal{F}^{-1}$
        $\mathcal{B} = \mathcal{F}^{-1}$
        $\mathcal{D} = \mathcal{U}^{-1}$

CHAPTER 4
Section 1.

4.1-1   The process $F^{-1}D^2F$ produces the permutation *(ufl, rbd)(fdl, bld)(fl, bd)(rd, ld)*.

4.1-2   List Notations:

urf → *ufl*
ufl → *ulb*
ulb → *fur*
ubr → *frd*
bdr → *ubr*
dfr → *bdr*
ur → *uf*
uf → *ul*
ul → *ub*
ub → *fr*
br → *ur*
dr → *br*
fr → *dr*

Cyclic Notation:
(urf, ufl, ulb)_(ubr, frd, drb)$_+$
(ur, uf, ul, ub, fr, dr, br)

4.1-3   a) *(ulb, ubr)$_+$(urf, frd)_(ub, ur, fr)*
b) *(ufl, ubr, bdr)*
c) *(uf, df, ub)*
d) *(ulb)_(ubr)$_+$(urf)$_+$(rbd)_(ub, br, ur)*

4.1-4   a) U
b) $RF^2R^{-1}B^2RF^2R^{-1}B^2$
c) $L^{-1}D^2LF^{-1}U^2FL^{-1}D^2LF^{-1}U^2F$

Section 2.

4.2-1   a) The permutation of both $F^2B^2L^2$ and $R^2L^2B^2F^2R^2$ is *(urf, ulb, drb, dlf)(ufl, dfr, dbl, ubr)(uf, df)(ul, dl)(ub, db)(fl, fr, bl, br)*.
b) The permutation of both $(F^2R^2)^2$ and $(R^2F^2)^4$ is *(ufl, dfr, ubr)(urf, drb, dlf)(fl, fr, br)*.
c) The permutation of both $(F^{-1}D^2FU^2)^2$ and $U^2RDR^{-1}U^2RD^{-1}R^{-1}$ is *(ufl, ubr, rbd)*.
d) The permutation of both $FR^{-1}F^{-1}R$ and $U^{-1}RUR^{-1}F^{-1}UFU^{-1}$ is *(ufl, bru)_ (urf, rdf)$_+$ (uf, rf, ru)*.

4.2-2 The permutation of FR is *(urf)₊ (ufl, rub, rbd, rdf, dlf)_ (uf, ru, rb, rd, rf, df, lf)* and the permutation of RF is *(dfr)₊ (ufl, rfu, rub, rbd, dlf)_ (uf, rf, ru, rb, rd, df, lf)*.

4.2-3 Yes. Equal permutations must be identical.
    i) Any process X has only one permutation.
    ii) Equality of permutations is symmetric.
    iii) Equality of permutations is transitive.

Section 3.

4.3-1 a) iv; b) vi; c) i; d) ii; e) iii; f) v.

4.3-2 a) False; b) True; c) False; d) True; e) False; f) True; g) True.

4.3-3 a) (B, A), (C, B, A), (D, C, B, A).
    b) It is the cycle in reverse order.
    c) (B, A) (E, D, C)
    d) Write each cycle of the inverse in its reverse order.

Section 4.

4.4-1 a) 4; b) 105; c) 63; d) 6; e) 2; f) 6.

4.4-2 $U^2LR^{-1}F^2L^{-1}R$

4.4-3 The largest order that any process on the cube can have is 1260. It is obtained by a twisted 3-cycle of corners, a 5-cycle of corners with the opposite twist, a flipped 2-cycle of edges, and another 2-cycle of edges and a 7-cycle of edges, one of which must also be flipped. The shortest example of such a process is $RF^2B^{-1}UB^{-1}$ which produces the permutation

$$(dfr, fdl, luf)_- (urf, bld, drb, ubr, bul)_+$$
$$(fu, fd, lu, br, dr, fl, fr)_+ (lb, ur)_+ (ub, db).$$

This process was presented by J.B. Butler.

Section 5.

4.5-1 $(FR^{-1})^9$

4.5-2 a) 7; b) 3; c) 2; d) 3; e) 12; f) 8; g) 3.

Section 6.

4.6-1 Only one permutation will do this. That is the permutation which flips all twelve edge pieces and otherwise leaves all pieces fixed. The shortest known process for doing this, attributed to M.B. Thistlethwaite, is

$$F^2B^2LF^2D^{-1}UR^2BL^2R^2F^{-1}L^2DU^{-1}B^2LFBUDRLFBUD$$

which takes 26 moves.

4.6-2 a) l; b) 6; c) 0, 2, 3, 4, 5, and 6.

4.6-3 $(ufl)_+(urf)_+(rdf)_+(uf)_+(rf)_+$

4.6-4 The permutation for both processes is
  a) *(ufl, urf)_ (rbd, rdf)_+ (uf, rd, rf)*
  b) *(ufl, rub)_+ (urf, frd)_ (uf, ru, rf)*
  c) *(urf, bru)_+ (fdl, frd)_ (ur, fd, fr)*

4.6-5 *(urf, ulb) (dlf, drb)*

4.6-6 $[X, Y]^{-1} = YXY^{-1}X^{-1} = [Y, X]$

Section 7.

4.7-1 The processes
$$B_1 = C_2A_2C_2^{-1}$$
$$B_2 = C_3A_3C_3^{-1}$$
$$B_3 = C_1A_1C_1^{-1}$$

where
$$C_1 = D^2R^2$$
$$C_2 = U^2F^2$$
$$C_3 = U^2F^2B^{-1}U^2F$$

For example
$$A_1 = LD^2L^{-1}ULD^2L^{-1}U^{-1}$$
$$A_2 = F[U,R]F^{-1} = FURU^{-1}R^{-1}F^{-1}$$
$$A_3 = BU^{-1}F^{-1}UB^{-1}U^2FU^{-1}F^{-1}U^2F$$

and
$$B_1 = U^2F^{-1}URU^{-1}R^{-1}FU^2 = U^2F^{-1}[U,R]FU^2$$
$$B_2 = U^2F^2B^{-1}U^2FBU^{-1}F^{-1}UB^{-1}U^2FU^{-1}FBF^2U^2$$
$$B_3 = D^2R^2LD^2L^{-1}ULD^2L^{-1}R^2D^2$$

4.7-2 $(XYX^{-1})^3 = (XYX^{-1})(XYX^{-1})(XYX^{-1})$
$$= XY(X^{-1}X)Y(X^{-1}X)YX^{-1}$$
$$= XY^3X^{-1}$$

4.7-3 a) $X = R$; b) $X = U^{-1}DF$; c) $X = UDFU^{-1}$; d) $X = U^2FU^2R$; e) $X = R^2D^{-1}$; f) $X = U^{-1}F^2R^{-1}F$; g) $X = RLD^{-1}B^{-1}$; h) $X = F$; i) $X = RFD^{-1}$; j) $X = RF^{-1}$.

4.7-4 $FRBF^{-1}LFD^{-1}F^{-1}L^{-1}FB^{-1}R^{-1}F^{-1}$

CHAPTER 6

Section 1.

6.1-1   a) Closure Law and Associative Law..
       b) All of them.
       c) Closure Law.
       d) None of them.
       e) Closure Law and Associative Law.
       f) Closure Law.
       g) Closure Law and Associative Law.

6.1-2   All of them.

6.1-3   a) These permutations form a group. This is the set of all permutations of Up-layer pieces.
       b) These form the group of all permutations of pieces of the Down layer and pieces of the middle layer between the Up and Down faces.
       c) These do not form a group. They do not satisfy either the Closure Law or the Identity Law.
       d) These form a group.

Section 2.

6.2-1   If x and y are combinations of n and m elements respectively from T and their inverses, then x·y can be expressed as a combination of $n+m$ elements from T and their inverses. Thus the Closure Law is satisfied. Since the Associative Law is satisfied for all elements of S, then it is satisfied for all elements of the subset $\langle T \rangle$ of S. Let s be any element of T. Then we have $s \cdot s^{-1} = 1$ in $\langle T \rangle$. Thus the Identity Law is satisfied. Finally if x is a finite combination of elements in T or their inverses,
$$x = s_1 s_2 \ldots s_n$$
where s_i is either in T or its inverse is in T, then
$$x^{-1} = s_n^{-1} \ldots s_2^{-1} s_1^{-1}$$
is a finite combination of elements in T or their inverses — that is, s_i^{-1} is either in T or its inverse is in T — Thus the Inverse Law is satisfied.

6.2-2   One element of a cyclic group can have order 2 if and only if the order, n, of the cyclic group is even.

6.2-3   a) 1; b) 1; c) 2; d) 2; e) 0; f) 2; g) 4.

6.2-4 a) $FRF^{-1}R^2$

b) i) $(FRF^{-1}R^2)^4$; ii) $(FRF^{-1}R^2)^3$; iii) $(FRF^{-1}R^2)^2$.

6.2-5 If x and y are in the cyclic group $G=\langle z\rangle$ then for some integers n and m we have

$$x=z^n$$
$$y=z^m.$$

Then we see that

$$x\cdot y=z^n\cdot z^m=z^{n+m}$$
$$=z^{m+n}=z^m\cdot z^n=y\cdot x.$$

6.2-6 The cyclic group $\langle X^k\rangle$ is identical to $\langle X\rangle$ for all integers, k, if and only if the order of X is a prime number.

Section 3.

6.3-1 Twelve

6.3-2 Six. Order 1-1; Order 2-7; Order 3-2; Order 6-2.

6.3-3 If $\langle U^2, R^2\rangle$ were cyclic, it would have to contain an element of order twelve. But in the previous exercise it was determined that no element had order greater than six.

6.3-4 Let $X=(U^2R^2)^2$ and let $Y=(U^2R^2)^3$. The permutation produced by X is

(ufl, ubr, dfr)(ulb, urf, drb)(ul, ur, dr)

and the permutation produced by Y is

(uf, ub)(fr, br).

These permutations are disjoint. Furthermore

$$U^2R^2=YX^{-1}$$

is in $\langle X, Y\rangle$. Therefore $\langle U^2R^2\rangle$ is contained in $\langle X, Y\rangle$. Also X and Y are in $\langle U^2R^2\rangle$ so $\langle X, Y\rangle$ is contained in $\langle U^2R^2\rangle$. Thus we have

$$\langle X, Y\rangle=\langle U^2R^2\rangle$$

which is a cyclic group.

Section 4.

6.4-1 $R^2L^2F^2B^2U^2D^2$

6.4-2 $RL^{-1}UD^{-1}BF^{-1}RL^{-1}$

6.4-3 $R^2L^2UD^{-1}F^2B^2UD^{-1}$

6.4-4 $F^2B^2U^{-1}DF^2B^2UD^{-1}$

6.4-5 a. The Slice-squared group is commutative. Thus every commutator reduces to I, the identity.

b. Eight.

c. Four.

Section 5.

6.5-1 $RUR^{-1}URU^2R^{-1}U^2R^{-1}U^{-1}RU^{-1}R^{-1}U^2RU^2$

6.5-2   Disregarding corners, edge pieces can be put in place by using conjugates of $[U, R]$. There is no need to flip any edges as their orientation never gets flipped in $\langle U, R \rangle$. Corners can now be put in their home locations by using conjugates of $[U, R]^3$. After the corners are put in place they can be correctly oriented to their home positions by using conjugates of the twist process of the proceeding exercise.

Section 6.

6.6-1   Since the Associative Law is satisfied for all of G, it is satisfied for the subset H of G. Let X be any element of H. Since H satisfied the Closure Law then X^k is in H for all positive integers, k. Since G is finite, the order, n, of X must be finite. Therefore, we have

$$X^n = I$$

is in H. Also we have

$$X^{n-1} = X^{-1}$$

in H. Therefore both the Identity Law and the Inverse Law are satisfied in H. Thus H is a group under the operation, *.

6.6-2   a) $(FBRL)^3$
       b) $(BFRL)^3U^2D^2$
       c) $FBUDR^{-1}L^{-1}FB$
       d) $(FBRL)^3(RLFB)^3$
       e) $(FBRL)^3(FBUD)^3$

6.6-3   Step 1; Restore the edge pieces *FL, RB, FD, RD* to their home positions. There are many ways to do this. For example, let us restore the pieces in the order they are listed above. To restore the *XY* piece, first place the *XY* piece in the *ub* location. Then move the *xy* location to the *rf* location — in the *rf* or *fr* position. Move the *XY* piece to either the *rf* or the *fr* position with either the move $F^{-1}U^2F$ or the move RUR^{-1} respectively. Then return the *XY* piece to its home position, *xy*. Step 2: Restore the remaining edge pieces using the method described in Step 4 of cube restoration in Chapters 3 and 5. Step 3: Place corners into their home locations using $[F, R]^3$, $[F, R^{-1}]^3$, $[U, R]^3$, $[U, R^{-1}]^3$, $[U, F]^3$, and $[U, F^{-1}]^3$. Step

4: Correct the orientation of corners using conjugates of the process given in Exercise 6.5-1.

6.6-4 $F^{-1}D^2U^2F^{-1}L^2D^2L^2R^2U^2R^2FD^2R^2F^2R^2F^2R^2L^2F^2L^2F^2U^2D^2R^2D^2F^{-1}$
$U^2D^2F^{-1}L^2R^2$

6.6-5 $R^{-1}L^2F^2B^2U^2F^2B^2R^2L^2D^2R^{-1}$

6.6-6 There are many pretty patterns. An extensive study of these patterns has been done by Richard Walker. Here are only a few of them.

    a) 4-Bars — $(R^2F^2L^2)^2$
    b) Crossbars — $R^2F^2L^2R^2F^2R^2$
    c) Double-cube —
       $BL^{-1}D^2LDF^{-1}D^2FD^{-1}B^{-1}F^{-1}RU^2R^{-1}U^{-1}BU^2B^{-1}UF$
    d) 2-X — $(F^2R^2)^3 \ (B^2L^2)^3$
    e) 4-U — $RLD^2R^{-1}L^{-1}FBD^2F^{-1}B^{-1}$
    f) 4-T — $FRBF^{-1}LFDF^{-1}L^{-1}FB^{-1}R^{-1}F^{-1}$

6.6-7   a) This is a subgroup.
    b) This is not a subgroup.
    c) This is not a subgroup.
    d) This is a subgroup.
    e) This is a subgroup.

Section 7.

6.7-1   a) $URLU^2R^{-1}L^{-1}URLU^2R^{-1}L^{-1}$
    b) $RL^{-1}FB^{-1}UD^{-1}R^{-1}U^{-1}DF^{-1}BR^{-1}LU$
    c) $L^2U^2R^{-1}L^{-1}URLU^2L^2B^{-1}FD^{-1}URDU^{-1}BF^{-1}$

6.7-2 Two thousand and forty-eight.

CHAPTER 7
Section 1.

7.1-1   *(uf, df) (ub, uf);*
    (ub, uf) (ub, df);
    (ub, uf) (uf, df) (ub, uf) (uf, df).

7.1-2   a) If P_1 and P_2 are odd permutations, then P_1 can be written as $2n+1$ pair exchanges and P_2 can be written as $2m+1$ pair exchanges. Then P_1P_2 can be written as $2n+1$ pair exchanges followed by $2m+1$ pair exchanges for a total of $2(n+m+1)$ pair exchanges. Since $2(n+m+1)$ is even then P_1P_2 is an even permutation.

    b) Even
    c) Odd

7.1-3 $F^{-1}UBU^{-1}FU^2B^{-1}UBU^2B^{-1}$

7.1-4   Let G be a group of permutations and let X be an odd permutation in G. Let A be the subset of even permutations in G. Then for each Y in A, XY is an odd permutation in G. Thus the number of odd permutations is equal to or greater than the number of even permutations. Similarly, if B is the subset of odd permutations in G, then for each Z in B, XZ is an even permutation in G. Thus the number of odd permutations cannot be greater than the number of even permutations. They must be equal. Thus, exactly half the permutations in G must be odd.

Section 2.

7.2-1   Since all edges have been restored, the permutation of edge pieces is even. Therefore the permutation of corner pieces is also even. Five corners remain to be restored. The even permutations of five elements are of the form either of a 5-cycle or of a 3-cycle or of two 2-cycles or the identity. Thus there are *four* possible cycle structures.

7.2-2   Any commutator produces an even permutation of edge pieces.

7.2-3   In the Slice group, the corner pieces can be used as a reference for the orientation of the other pieces of the cube. Thus, with the corner positions fixed, each quarter turn of a slice produces an odd permutation of edge pieces and a single exchange of an opposing center piece pair. Since opposite center pieces always must remain opposite, they, as a pair, can be considered as a single piece with two possible orientations in any location. Both "spot" and "solid" patterns produce the identity permutations on edge pieces. To produce this identity then requires an even number of slice quarter turns, which in turn implies an even permutation of center piece pairs. There are exactly three even permutations of the three center-piece pairs.

    For each permutation of the center-piece pairs, we must determine how many orientations are possible. A center-piece pair has two orientations attained by flipping the piece end-for-end. The permutation of such a flip is a single pair exchange of the two center piece facelets. But again a quarter turn of a single slice produces an odd permutation of center-piece facelets. Thus a Slice group process producing an identity on the edge pieces must also produce an even permutation of center piece facelets. Thus, for a fixed permutation of center-piece pairs, the number of flips of center pieces must be even. This produces four possible orientations for the set of three center piece pairs. Thus the total number of permutations of center pieces produced by processes in the Slice group which leave both corner pieces and edge pieces fixed is twelve.

7.2-4 Any two disjoint pair exchanges can be expressed as two 3-cycles as follows:

$$(a, b)(c, d) = (a, c, b)(c, b, d).$$

Any two overlapping pair exchanges combine to form a single 3-cycle. Thus any even permutation of edges can be written as a sequence of 3-cycles. Any 3-cycle of edges can be produced by a conjugate, $X[F, R]X^{-1}$, where the process, X, moves the three edge locations to be cycled into the locations $\{uf, rf, rd\}$. Thus, by disregarding corners, any even permutation of edges can be produced by conjugates of $[F, R]$.

To put corners in place without disturbing edges we use

$$(X[F, R]X^{-1})^3 = X[F, R]^3 X^{-1}$$

which produces two disjoint corner pair exchanges. Any 3-cycle of corners can be produced by

$$X[F, R]^3 U[F, R]^3 U^{-1} X^{-1} = (X[F, R]^3 X^{-1})(XU[F, R]^3 U^{-1} X^{-1})$$

where X is a process which moves the three corners to be cycled into the locations $\{urf, flu, bru\}$. Thus any even permutations of corners can be produced by conjugates of $[F, R]$ without disturbing edges.

Proper corner twists can be achieved by choosing the conjugate for the previous corner permutations carefully. However, they can also be produced by conjugates of

$$[F, R]^2 L[F, R]^{-2} L^{-1} = [F, R]^2 (L[F, R]L^{-1})^4.$$

Section 3.

7.3-1 Three

7.3-2 Four

7.3-3 Each generator of the Squares group leaves every piece sane. Thus every finite combination of squares will leave every piece sane.

7.3-4 Consider the Two-generator group, $\langle U, R \rangle$. Each of the generators leaves all edge pieces sane. Thus no finite combination of moves of the generators can cause any edge piece to be flipped.

7.3-5 8-twist: $(LR^2F^2B^{-1})^4$

4-flip: $F^2B^2LF^2UD^{-1}R^2BR^2L^2F^{-1}L^2U^{-1}DB^2R^{-1}$

8-flip: $(RLFBUD)^2$

Section 4.

7.4-1 For $n = 1$, the only possible permutation of a single object in a single location is the identity. The number of permutations therefore is $1! = 1$.

Assume that the number of permutations of $(n-1)$ objects in $(n-1)$ locations is $(n-1)! = (n-1)(n-2) \cdots 3 \cdot 2 \cdot 1$. To find the number of permutations — that is, rearrangements — of n objects in n locations choose one location to be the first location. Observe that there are n possible objects that could be placed in the first location. For each choice for the first location, there are $(n-1)$ objects to be arranged in the remaining $(n-1)$ locations. By our assumption, there are $(n-1)!$ possible permutations of these $(n-1)$ objects in these $(n-1)$ locations, for each choice of an object for the first location. Combining the number of permutations for all n possible choices for the first location we get $n \cdot (n-1)! = n!$ possible permutations of n objects in n locations.

By mathematical induction, the assertion is shown to be true for all values of n.

7.4-2 $44,089,920 = 8! 3^8/6$

This counts all even permutations of corner locations times the number of corner twists to produce "sanity" modulo 3.

Section 5.

7.5-1 The brute-force way to solve this exercise is to list all permutations of corner locations formed by rotating the Up and Right faces. When further turns of the Up and Right faces only produce permutations which have been produced by processes with fewer moves then all permutations have been generated and you can count them. If you do this you will find that the number is 120. To prove this directly is tricky. Here is one way.

Let us number the corners as follows: *ufl* $= 1$, *ulb* $= 2$, *ubr* $= 3$, *urf* $= 4$, *dfr* $= 5$, and *drb* $= 6$. Then the permutations of the process U, in cyclic notation is

$$(1, 2, 3, 4)$$
and the permutation of the process R in cyclic notation is
$$(6, 5, 4, 3).$$
Consider all possible pairs of the corners, ij, without regard for order. We have
$$15 = 6 \cdot 5/2$$
such pairs with i different from j. We consider ij and ji to be the same pair so we will always write it with i<j. We group these pairs into five sets of three pairs each, as follows:
$$A = \{12, 35, 46\}$$
$$B = \{16, 23, 45\}$$
$$C = \{15, 26, 34\}$$
$$D = \{14, 25, 36\}$$
$$E = \{13, 24, 56\}$$
As we did with the pairs, we again consider these as unordered sets, *not* as ordered triples.

Now observe that the process U moves the set A onto the set B. That is
$$12 \rightarrow 23$$
$$35 \rightarrow 45$$
$$46 \rightarrow 16$$
Further examination shows that the process U actually permutes the sets A, B, C, and D so that we have the permutation
$$(A, B, C, D).$$
The process R causes the permutation
$$(B, C, D, E).$$
Thus we can see that every process in $\langle U, R \rangle$ produces a permutation of the sets A, B, C, D, and E. Furthermore, for every X and Y in $\langle U, R \rangle$, the permutation produced by XY is the same as the permutation produced by X followed by the permutation produced by Y.

We now must show that if X and Y are two processes in $\langle U, R \rangle$ which produce the same permutation of A, B, C, D, and E then they must be equivalent processes in $\langle U, R \rangle$. If X and Y are any two processes in $\langle U, R \rangle$ then for some process Z we have
$$XZ = Y.$$
If X and Y produce the same permutation of A, B, C, D, and E then Z must produce the identity permutation on A, B, C, D, and E. We now show that if Z produces the identity on A, B, C, D, and E then it must produce the identity on the corner locations in $\langle U, R \rangle$. To do that, we suppose the ith corner moved to the jth location. For one of

the sets A, B, C, D, or E, we have ij as one of the pairs — disregarding order. Therefore it must move to itself and we see that the jth corner must move to the ith location. But in a second of the sets, A, B, C, D, and E, we have ik and jm while in a third we have jk and in with $n \neq m$. But in the second set we must have the kth corner exchanging with the mth corner, while in the third set the kth corner must exchange with the nth corner. This contradiction shows that our assumption that some i^{th} corner moved to the jth location must be false. Thus Z must produce the identity on the corner locations in $\langle U, R \rangle$. Thus X and Y both produce the same permutation on the corners of $\langle U, R \rangle$.

This shows that the number of permutations of corner locations produced by processes in $\langle U, R \rangle$ is not greater than the number of permutations of A, B, C, D, and E. We saw in Exercise 7.4-1 that the number of permutations of five objects in five locations is $5! = 120$. Thus there are no more than 120 permutations of corner locations.

To show that there are at least 120 such permutations, we first see that we can place any of the six corners in the *ulb* location. Any of the remaining five can be placed in the *ufl* corner, if it is not already there, by a process of the form $U^{-1}RU$. Then any of the remaining four can be placed in the *urf* corner by rotating the Right face. This produces at least $120 = 6 \cdot 5 \cdot 4$ different permutations.

Thus we see that the total number of permutations of corner locations produced by processes in $\langle U, R \rangle$ is 120.

7.5-2   We first show that b) and c) are equivalent.

We showed in the previous exercise that any three corner locations can be moved by a process in $\langle U, R \rangle$ to the *ulb, ufl,* and *urf* locations. Hence if any single 3-cycle of corner locations were produced by a process in $\langle U, R \rangle$ then a conjugate of that process would produce a 3-cycle of the locations, *ulb, ufl,* and *urf*. Then a quarter turn of the Up face would restore two of those corners to their home locations and produce a 2-cycle of the other two Up-face corner locations. Thus we have shown that c) implies b).

Similarly if any 2-cycle of corner locations were produced by a process in $\langle U, R \rangle$ then one conjugate of that process would produce a 2-cycle of the locations, *ulb* and *ufl*, and another conjugate would produce a 2-cycle of the locations, *ufl* and *urf*. Letting one of these conjugates follow the other would then produce a 3-cycle of the locations, *ulb, ufl,* and *urf.* Thus we have shown that b) implies c).

We now show that a) implies b) — again by showing that NOT b) implies NOT a). We have observed that any three corner locations can be moved by a process in $\langle U, R \rangle$ to the locations *ulb, ufl,* and *urf.* Thus there are at least 120 permutations. The same argument with the cube reoriented so as to exchange the Up face and the Right face and then returned, shows that any three corner locations can be moved to the locations *rdf, rbd,* and *rub.* Thus if there is a process in $\langle U, R \rangle$ which produces a single 3-cycle of corner locations, then a conjugate of that process will produce a 3-cycle of the locations, *rdf, rbd,* and *rub,* leaving the other locations in place. Thus for each of the 120 ways that corners can be placed in the *ulb, ufl,* and *urf* locations there would be at least three ways to place corners in the *rdf, rbd,* and *rub* locations. Thus there would be at least 360 permutations of corner locations. Hence we see that a) implies b).

To show that b) implies a) we will show that NOT a) implies either NOT b) or NOT c). We have seen that there are 120 ways to move the six corner locations of $\langle U, R \rangle$ to the locations, *ulb, ufl,* and *urf.* If there are more than 120 permutations of corner locations in $\langle U, R \rangle$ then there are two processes, X and Y, in $\langle U, R \rangle$ which move the same corner locations to *ulb, ufl,* and *urf* but move different corner locations to *rdf, rbd,* and *rub.* Then the process XY^{-1} will leave the three corners which were moved by X to *ulb, ufl,* and *urf* in their home locations. However the other three corners — which X moved to *rdf, rbd,* and *rub* — are not left in their home locations by XY^{-1}. But any permutation other than the identity of three objects in three locations is either a single 3-cycle or a single 2-cycle. Thus we have either NOT b) or NOT c). This completes the proof.

7.5-3 14,400

Section 6.

7.6-1 a) Any even permutation of corners can be produced by using conjugates of
$$(UR^2U^{-1}L^2)^2.$$
Any odd permutation of corners can be reached from any other odd permutation in the same way. Thus any permuta-

tion of corners is possible. There are 8! such permutations of the eight corners.

Any permutation of the eight edge pieces on the Up and Down faces can be produced without affecting the corners by conjugates of the process

$$(U^2R^2)^3.$$

For each of the corner permutations, there are 8! such permutations of the Up and Down-face edges.

Any even permutation of the middle-layer edge pieces can be produced without effecting the Up and Down-face pieces by the processes $R^2UD^{-1}F^2U^{-1}D$ and $R^2U^{-1}DB^2UD^{-1}$. There are 24 permutations of the four middle-layer edges, half of which are even and half of which are odd. If the permutations of the Up and Down-face locations is odd or even then the permutations of the middle-layer edges must also be odd or even respectively. Since, in the middle-layer edge permutations every odd permutation can be reached from any other odd permutation by an even permutation, then for each permutation of the Up and Down-face pieces there are 12 permutations of the middle-layer edges.

Thus the total number of permutations in $\langle U, D, R^2, L^2, F^2, B^2 \rangle$ is $(8!)^2 \cdot 12$.

b) The process $U^{-1}DR^2UD^{-1}$ in $\langle U, D, R^2 \rangle$ has the same affect on the Up and Down faces as the process F^2 had in $\langle U, D, R^2, L^2, F^2, B^2 \rangle$. In a similar way we can produce the same effect as L^2 and B^2 on the Up and Down faces. Thus we have the same permutations of the Up and Down-face pieces in $\langle U, D, R^2 \rangle$ as in $\langle U, D, R^2, L^2, F^2, B^2 \rangle$. However the only possible permutations of the middle layer edges are the identity and *(fr, br)*. Since an odd permutation of Up and Down-face pieces must go with an odd permutation of the middle layer and an even with an even, there is only one middle-layer permutation for each Up and Down-face permutation. Thus the order of $\langle U, D, R^2 \rangle$ is $(8!)^2$.

7.6-2   a) $2^{29} \cdot 3^{15} \cdot 5^3 \cdot 7^2 \cdot 11 = 519,024,039,293,878,272,000$
       b) $2^{41} \cdot 3^{15} \cdot 5^3 \cdot 7^2 \cdot 11 = 2,125,922,462,947,725,402,112,000$

7.6-3   a) $|U^2, R^2| = 12 = 2^2 \cdot 3$
       b) $|U, R, D| = 159,993,501,696,000 = 2^{14} \cdot 3^{13} \cdot 5^3 \cdot 7$
       c) $|U^2, R^2, D^2| = 96 = 2^5 \cdot 3$

    d) $|U, R, F| = 170,659,735,142,400 = 2^{18} \cdot 3^{12} \cdot 5^2 \cdot 7^2$
    e) $|U^2, R^2, F^2| = 2,592 = 2^5 \cdot 3^4$
    f) $|U^2, R^2, F^2, B^2| = 165,888 = 2^{11} \cdot 3^4$
    g) $|U^2, R^2, F^2, B^2, L^2| = 663,552 = 2^{13} \cdot 3^4$
    h) $|U^2, R^2, F^2, B^2, L^2, D^2| = 663,552 = 2^{13} \cdot 3^4$
    i) $|U, R^2, L^2, F^2, B^2| = 6,502,809,600 = 2^{16} \cdot 3^4 \cdot 5^2 \cdot 7^2$

7.6-4 a) If x and y are in C then for every g in G we have
$$(x \cdot y) \cdot g = x \cdot (y \cdot g) = (g \cdot y) \cdot x = g \cdot (y \cdot x) = g \cdot (x \cdot y).$$
Hence $(x \cdot y)$ is in C and the Closure Law is satisfied. The Associative Law is satisfied for all subsets of G. The identity is in C since the identity element commutes with every element of G by the Identity Law for G. To show that the inverse of each element in C is also in C, consider any element x in C. We have
$$x^{-1} \cdot g = (g^{-1} \cdot x)^{-1} = (x \cdot g^{-1})^{-1} = g \cdot x^{-1}$$
and therefore x^{-1} is in C. Hence C is a subgroup of G.
b) There are only two elements in C, the identity and the permutation which flips every edge piece of the cube in place and leaves all the corners fixed (See Exercise 4.6-1).

CHAPTER 8

Section 1.

8.1-1 $\langle U, R^2UD^{-1}F^2U^{-1}D, RDR^{-1}[U, L]RD^{-1}R, FR^{-1}F^{-1}RF^{-1}U^{-1}F \rangle$

8.1-2 $\langle R^2UD^{-1}F^2U^{-1}D, F^2UD^{-1}L^2U^{-1}D,$
$R^{-1}FD^{-1}RF^{-1}DU^{-1}RB^{-1}DR^{-1}BUD^{-1} \rangle$

8.1-3 Each of the generators leaves all the pieces *sane*.

8.1-4 The Magic Domino can perform exactly the processes U, D, R^2, L^2, F^2, and B^2. The permutation of the Up and Down faces produced by any process on the cube is identical to the permutation produced by that process on the Magic Domino. Thistlethwaite has shown that any permutation in $\langle U, D, R^2, L^2, F^2, B^2 \rangle$ can be restored in no more than 32 moves — see Table 8.1-1. This number might be able to be lowered by ignoring the middle layer but would certainly not be increased.

8.1-5 a) $G_1 = \langle U, R^{-1}UR, BUB^{-1}, LUL^{-1}, FUF^{-1}, RUR^{-1} \rangle$
    b) $G_2 = \langle U, R^{-1}UR, BUB^{-1}, DLUL^{-1}D^{-1}, D^2FUF^{-1}D^2,$
$D^{-1}RUR^{-1}D \rangle$

   c) $G_3 = \langle U, R^{-1}UR, BUB^{-1} \rangle$

   d) $G_4 = \langle [LD^2L^{-1}, U^i], [F^{-1}D^2F, U^i] \rangle$ for $i = 0, 1, 2, 3$.

   e) $G_5 = \langle [LD^2L^{-1}F^{-1}D^2F, U^i], R[LD^2L^{-1}F^{-1}D^2F, U^2]R^{-1} \rangle$

8.1-6 True

Section 2.

8.2-1 First, every element, x, in G is in at least one right coset, in particular, in the coset Hx.

   Let z be in both Hx and Hy. Then for some h_1 and h_2 in H we have

$$h_1 x = z = h_2 y.$$

Now for every $h_3 x$ in Hx, we have

$$h_3 x = h_3 h_1^{-1} h_1 x = h_3 h_1^{-1} h_2 y$$

and thus $h_3 x$ is in Hy. Similarly for every $h_4 y$ in Hy, we have

$$h_4 y = h_4 h_2^{-1} h_2 y = h_4 h_2^{-1} h_1 x$$

and $h_4 y$ is in Hx. Thus we see that if Hx and Hy are not disjoint then they are identical.

8.2-2 If H is a finite subgroup of the group G, then, for each right coset, Hx, of H, the number of elements in Hx is the same as the number of elements in H. To see this we need only pair-up the elements

$$h \rightarrow hx$$

and observe that $h_1 x = h_2 x$ if and only if $h_1 = h_2$. Thus every coset of H has the same number of elements. We saw in the previous exercise that all the cosets were disjoint. Therefore, if G is finite, the number of elements in G is equal to the number of elements in H times the number of cosets of H in G. Thus the order of H divides the order of G.

8.2-3 a) $9,072 = 8 \cdot 7 \cdot 6 \cdot 3^3$

      b) $2,688 = 8 \cdot 7 \cdot 6 \cdot 2^3$

      c) $960 = 5 \cdot 4 \cdot 3 \cdot 2^4$

      d) $60 = 5 \cdot 4 \cdot 3$

      e) $81 = 3^4$

      f) $9,270,405,365,760 = 2^{19} \cdot 3^8 \cdot 5 \cdot 7^2 \cdot 11$

      g) $466,560 = 2^8 \cdot 3^6 \cdot 5^2$

8.2-4 The index of H_1 in H_0 is $2,048 = 2^{11}$.

      The index of H_2 in H_1 is $1,082,565 = 3^9 \cdot 5 \cdot 11$.

      The index of H_3 in H_2 is $29,400 = 2^3 \cdot 3 \cdot 5^2 \cdot 7^2$.

      The index of H_4 in H_3 is $663,552 = 2^{13} \cdot 3^3$.

8.2-5 a) $6,123,600 = 2^4 \cdot 3^7 \cdot 5^2 \cdot 7$
       b) 2
       c) 2
       d) $88,179,840 = 2^7 \cdot 3^9 \cdot 5 \cdot 7$
       e) $2,048 = 2$

Section 3.

8.3-1 If H is a normal subgroup of G and x is an element in G then xH=Hx. Thus for every h_1 in H there is an h_2 in H such that

$$x \ast h_1 = h_2 \ast x.$$

Therefore we have

$$x \ast h_1 \ast x^{-1} = h_2,$$

and $xHx^{-1} \subset H$. For every h_1 in H there is also an h_3 in H such that

$$x^1 \ast h_1 = h_3 \ast x^{-1}$$

or equivalently

$$h_1 = x \ast h_3 \ast x^{-1}$$

Thus we also have $H \subset xHx^{-1}$. The two results together show that if H is a normal subgroup of G then $H = xHx^{-1}$.

Suppose we have $xHx^{-1} = H$ for every x in G. Then for every h_1 in H there are h_2 and h_3 in H such that

$$x \ast h_1 \ast x^{-1} = h_2$$

and

$$h_1 = x \ast h_3 \ast x^{-1}$$

Equivalently we have

$$x \ast h_1 = h_2 \ast x$$

and

$$h_1 \ast x = x \ast h_2.$$

This shows that

$$xH \subset Hy$$

and

$$Hx \subset xH.$$

Thus we have xH=Hx and H is a normal subgroup of G.

8.3-2 To show that the Closure Law is satisfied for the operation which combines left cosets, xH and yH, of H we must show that $xH \ast yH = \{z : z = (x \ast h_1) \ast (y \ast h_2) \text{ for some } h_1 \text{ and } h_2 \text{ in } H\}$ is a left coset of H. Since Hy=yH we have h_3 in H such that

$$h_1 \ast y = y \ast h_3$$

and
$$(x*h_1)*(y*h_2)=x*(h_1*y)*h_2=x*(y*h_3)*h_2=(x*y)*(h_3*h_2)$$
Thus if z is in xH*yH then z is in (x*y)H. Also if z is in (x*y)H then
$$z=(x*y)h=(x*1)*(y*h)$$
and we see that z is in xH*yH. Thus
$$xH*yH=(x*y)H.$$
From this equality all the other group criteria follow trivially. Associativity follows directly from associativity in the group G. The Identity Law is satisfied by the identity coset,
$$1H=H,$$
since
$$xH*H=(x*1)H=xH=(1*x)H=H*xH.$$
The equality
$$xH*x^{-1}H=(xx^{-1})H=H=(x^{-1}*x)H=x^{-1}H*xH$$
shows that the Inverse Law is satisfied. Hence the set of left cosets, G/H, is a group.

8.3-3   Let X be any element of G and E_1 be any element of A_e. Then X can be decomposed into two disjoint permutations X_c and X_e where X_c is the permutation caused by X on the corners and X_e is the permutation caused by X on the edges. No claim is made that any *process* exists on the cube which will produce either X_c or X_e. In fact, if they are odd permutations, no such process will exist. Thus, we have
$$X_cX_e=X.$$
Now consider
$$XE_1X^{-1}=(X_cX_e)E_1(X_e^{-1}X_c^{-1})=X_c(X_eE_1X_e^{-1})X_c^{-1}.$$
Since X_e and X_e^{-1} are either both odd or both even and E_1 is even then $X_eE_1X_e^{-1}$ is even and in A_e. Since X_c and $X_eE_1X_e^{-1}$ are disjoint permutations, then we have
$$X_c(X_eE_1X_e^{-1})=(X_eE_1X_e^{-1})X_c$$
and thus we see that
$$X_c(X_eE_1X_e^{-1})X_c^{-1}=X_eE_1X_e^{-1}$$
is in A_e. Thus, using Exercise 8.3-1, we have shown that A_e is a normal subgroup of G.

8.3-4   Let C be any element of A_c. We will show that the mapping

$$C \rightarrow CA_e$$

is an isomorphism between A_c and A/A_e.

To show that this mapping is one-to-one, suppose we have

$$C_1 A_e = C_2 A_e.$$

Then we have $(C_1 C_2^{-1})A_e = A_e$ and thus $C_1 C_2^{-1}$ is in A_e. But C_1 and C_2^{-1} both leave all edge pieces fixed and the only element in A_e which leaves all edge pieces fixed is the identity. Thus we have

$$C_1 C_2^{-1} = 1$$

and

$$C_1 = C_2.$$

Therefore the mapping is one-to-one. To show that A_c is mapped *onto* A/A_e, consider any element XA_e. Since X is in A it produces both an even permutation of corners and an even permutation of edges. Let X_e be the permutation of edges produced by X but leaving all the corners fixed. This exists since the permutations of edges and corners are disjoint. Also, since X_e is even, X_e is in A_e, and X_e^{-1} is also in A_e. Thus we have XX_e^{-1} in XA_e, and $XA_e = (XX_e^{-1})A_e$. But XX_e^{-1} is an identity on all the edge pieces and hence is in A_c. Thus the XX_e^{-1} is mapped onto $(XX_e^{-1})A_e = XA_e$. Therefore the mapping is onto all of A/A_e. We showed in the previous exercise that A_e was a normal subgroup and, in Exercise 8.3-2 that for any normal subgroup such as A_e we have

$$(C_1 A_e)(C_2 A_e) = (C_1 C_2)A_e.$$

Hence the mapping from A_c onto A/A_e is an isomorphism.

8.3-5   a) $\langle F^2 R^2 \rangle$
        b) $\langle (FR)^3 \rangle$
        c) $\langle R^2 L^2 UD^{-1} F^2 B^2 UD^{-1}, RL^{-1} FB^{-1} UD^{-1} RL^{-1} \rangle$

8.3-6   For any X in G, define the mapping, f, from G onto G/A_c,

$$f(X) = XA_c.$$

For any X and Y in G, since A_c is a normal subgroup of G — see Exercise 8.3-2 — we have

$$f(X) \cdot f(Y) = XA_c \cdot YA_c = XYA_c = f(XY).$$

Thus f is a homomorphic mapping of G onto G/A_c.

8.3-7   Each permutation X in H_2 can be decomposed into two dis-joint permutations X_{UD} and X_m, where X_{UD} is the permutation of the Up and Down-face pieces and X_m is the permutation of the pieces in the middle layer between the Up and Down faces. Define a 1-1 mapping, f, between the Up and Down-face locations of the cube and the locations on the Magic Domino such that each location on the cube corresponds to the same location on the domino — for example, $f(ulb) = ulb$ on the domino. Then the mapping

$$X \rightarrow f^{-1}X_{UD}f$$

is a homomorphism of H_2 onto the Magic Domino.

8.3-8   a) True; b) True; c) False; d) True.

8.3-9   a) No; b) Yes; c) Yes, Index 4.

8.3-10   a) True; b) False; c) True; d) True.

INDEX

Also read what *Scientific American* calls "the definitive treatise"
on Rubik's Magic Cube:

NOTES ON RUBIK'S MAGIC CUBE
by David Singmaster

Notes On Rubik's Magic Cube was written over the period from 1978
to 1981 when the remarkable properties of the cube were first being
appreciated and our understanding codified. As a result, the book gives
the reader the excitement of participating in the development of a new
field. *Notes* started as a booklet of 30 pages, to which addenda were
added successively. Addendum Number Five is almost as long as the
original work itself. The book is chock-full of amusing anecdotes,
happy phrases, and colorful personalities and is a repository of lore
about the cube.

Notes also presents the Singmaster notation system, the basic concepts of permutations and group theory, a treatment of simpler subgroups of the cube, some open-ended problems, a 200-move solution
for restoring the cube, and a catalog of useful processes.

The cube originally was invented by Ernö Rubik to help students of
industrial engineering develop their three-dimensional abilities. Frey
and Singmaster's *Handbook Of Cubik Math* helps readers develop their
mathematical abilities, using the cube. David Singmaster's *Notes On
Rubik's Magic Cube* is a backup reference tool on the cube, with a
wealth of additional material for readers whose sophistication may
vary over a wide range.

Hardcover edition ISBN 0-89490-057-9
Paperback edition ISBN 0-89490-043-9

Write for prices.

Available from
ENSLOW PUBLISHERS
Bloy Street and Ramsey Avenue
Box 777
Hillside, New Jersey 07205